IMAGES
of America

CAMP RUCKER
DURING WORLD WAR II

An M4 Sherman tank of the 746th Tank Battalion rumbles through the scrub brush and pine trees of Camp Rucker. Activated at Camp Rucker in August of 1942, the 746th Tank Battalion was destined to land on Utah Beach in Normandy on D-Day and then fight its way across France and Belgium and into Germany by the end of the war. (Courtesy of AAM.)

IMAGES
of America

CAMP RUCKER
DURING WORLD WAR II

James L. Noles Jr.

ARCADIA
PUBLISHING

Published by Arcadia Publishing
Charleston, South Carolina

Library of Congress Catalog Card Number: 2002112516

For all general information contact Arcadia Publishing at:
Telephone 843-853-2070
Fax 843-853-0044
E-Mail sales@arcadiapublishing.com
For customer service and orders:
Toll-Free 1-888-313-2665

Visit us on the Internet at www.arcadiapublishing.com

Col. Lloyd S. Spooner, who commanded Camp Rucker from August 31, 1943 to July 12, 1944, realized that much of his camp's effectiveness depended upon the continued good will of the citizens of Alabama. Here, he hosts a delegation from Montgomery for a tour of the camp. (Courtesy of AAM.)

CONTENTS

The flags of the 81st Infantry Division ("Wildcats") pass in review at Camp Rucker in 1943. Activated on June 15, 1942, this division was the first major unit to be formed at the camp. After deploying to the Pacific in 1944, the division's GIs would see combat in Peleliu, Ulithi, Pulo Anna, and the Philippines. (Courtesy of AAM.)

ACKNOWLEDGMENTS

Great thanks are due at the outset to Dr. Jim Williams of Fort Rucker's Army Aviation Museum. If Dr. Williams had not been willing to share the museum's collection of World War II–era photographs with me, then this book would have simply been impossible. Photographs from the Army Aviation Museum are credited "Courtesy of AAM." I also greatly appreciated Ray Roberts, a veteran of the 766th Ordnance Company, for sharing some of his wartime photographs with me. Credit is also due to Judge Val McGee, whose book *The Origins of Fort Rucker* provides an essential reference to the development of Camp, and later Fort Rucker. For more information on Alabama during World War II, one should read Dr. Allen Cronenberg's *Forth to the Mighty Conflict: Alabama and World War II.*

Camp Rucker During World War II is, at its heart, a book about soldiers. For that reason, I am taking the liberty of dedicating it to my father, James L. Noles, former brigadier general, U.S. Army of Florence, Alabama. If there was ever an officer who loved his soldiers more than my father, I certainly never met him.

Perhaps most importantly, as with all of my writing projects, I continue to be profoundly grateful to my lovely wife Elizabeth and my son James for their willingness to tolerate my evenings and weekends at the computer keyboard. Thank you both.

INTRODUCTION

As July of 1990 drew to a close, Ken Hawley, Mike Mazur, Don Sparaco, and I—four brand-new U.S. Army second lieutenants—drove south down U.S. Highway 231. We were destined for Fort Rucker, nestled deep in the southeastern corner of Alabama. With the bronze bars on our shoulders less than two months old, our thoughts were focused solely on our future in the Army.

Fort Rucker, however, represented far more than simply the destination on our own immediate personal horizons. The post's history spanned nearly five decades. As the four of us passed through its gates for the first time, we took our places in a long green line. Its origins lay in the troubles of the Great Depression and the outbreak of World War II.

Throughout much of the Great Depression, local United States Congressman Henry B. Steagall had lobbied hard for federal support for the struggling townspeople and farmers of his southeast Alabama district. A federal land purchase program, relieving farmers of marginal and substandard cropland, provided some assistance, but an Army base in his district eventually seemed to be the best bet. On July 16, 1941, the War Department announced that it would build a training camp for 30,000 soldiers just outside of Ozark, Alabama. At first, the camp was known simply as the "Ozark Triangular Division Camp," taking its name from the organizational structure of American divisions at the time. Such divisions contained three regiments, each of which in turn contained three battalions—hence the designation "triangular."

Construction commenced at the camp on July 15, 1942. By the end of the month, the War Department announced that the new camp would be named Camp Rucker, after Confederate cavalry officer Col. Edmund W. Rucker. Rucker had lost his left arm in a horseback melee during the Battle of Nashville. Following the war, Rucker had moved to Birmingham, where he had become one of the young city's leading industrial pioneers.

In April of 1942, while the J.A. Jones Construction Company was still frantically hammering together the camp's 1,500 buildings, Col. Frederick Manley set up his headquarters at the camp as the new garrison commander. Brig. Gen. Gustav H. Franke, commanding the trainees of the brand-new 81st Infantry Division but succeeded before the end of the summer by Brig. Gen. Paul J. Mueller, also settled into his new headquarters that month. On May 1, 1942, Colonel Manley officially activated Camp Rucker.

Overseas, May 1942 was a grim month for the Allies' war effort. In its first week, the last American troops in the Philippines surrendered the island fortress of Corregidor to the Japanese and began over three years of captivity. Meanwhile, in Russia, the Germans were beginning their summer Crimean offensive, while in North Africa Rommel was marshalling his *Afrika Korps* for an offensive that would eventually capture the British citadel of Tobruk. In England, German bombers continued to pound civilian targets in a ruthless campaign of terror bombing.

Born in such desperate times, Camp Rucker was destined to play a key role in training America's young men (and women) for World War II. A steady stream of units passed through Camp Rucker's gates, training for combat in Europe and the Pacific. They included large units like the 81st Infantry Division (nicknamed "Wildcats"), the 35th Infantry Division ("Sante Fe"), the 98th Infantry Division ("Iroquois"), and the 66th Infantry Division ("Panthers"). Dozens of smaller units trained in their shadow: the 640th and the 628th Tank Destroyer Battalions, the 746th Tank Battalion, the 37th Medical Ambulance Battalion, the 336th Engineer Combat Battalion, the 84th Chemical Mortar Battalion, the 14th Chemical Maintenance Company, and many more. By February of 1945, new recruits were also training as individual replacements at the camp's Infantry Training Replacement Center, destined to fill gaps in units decimated by hard fighting in Europe and the Pacific.

Meanwhile, faced with preparing its young citizen soldiers for an unfamiliar life in the Army, the War Department, on August 10, 1944, published an indoctrination pamphlet entitled simply *Army Life*. *Army Life* explained to its readers that its words "reach you as your Army life is beginning. They have been written by men who have gone through what you are now experiencing. They are intended to help you become the finest soldier in the world by answering many of the questions on your mind, by putting you at ease in your new surroundings." One can easily imagine the freshly shorn buck privates of Camp Rucker thumbing through its pages, trying to glean some hint as to what awaited them during their basic training. Quoted passages from *Army Life* accompany the photographs contained in *Camp Rucker During World War II*. Hopefully, these passages help to breathe further life into the scenes portrayed by the book's photographs.

With Japan's capitulation in August of 1945, Camp Rucker's wartime mission drew to a close. It would spring back to life to train more units for combat service during the Korean War. Then, on October 26, 1955, Camp Rucker achieved the coveted status of a permanent installation when the Department of Defense designated it "Fort Rucker." As the Army Aviation Training Center, it would train, and continues to train today, thousands more men and women on its grounds and in the skies above. The veterans of World War II, however, paved their paths through Fort Rucker. *Camp Rucker During World War II* provides a brief glimpse at those veterans' story.

One

ONE CAMP,
MANY LEADERS

Although Col. Frederick Manley formally activated Camp Rucker on May 1, 1942, he and his staff did not move into the camp headquarters until its completion the following month. This photograph shows the headquarters building in August of 1943. "In the Army," *Army Life* explained, "all orders are issued 'through channels,' or following the 'chain of command.' This extends from the bottom to the top . . . . This system has a sound basis. An organization as tremendous as the Army would bog down without such a system." (Courtesy of AAM.)

Brig. Gen. Frederick Manley shakes hands with W.A. Steadman of the Alabama State Chamber of Commerce on March 5, 1943. Brigadier General Manley commanded Camp Rucker for 16 months until his retirement on August 31, 1943. A 1905 graduate of West Point and a veteran of World War I, Manley's maturity and organizational skills played a key role in the initial success of the infant camp. (Courtesy of AAM.)

A delegation of "prominent city and civic officials" of nearby Troy, Alabama, paid a visit to Camp Rucker on April 27, 1944. In the middle of the group stands Col. Lloyd S. Spooner, who followed General Manley as the camp's commanding officer. (Courtesy of AAM.)

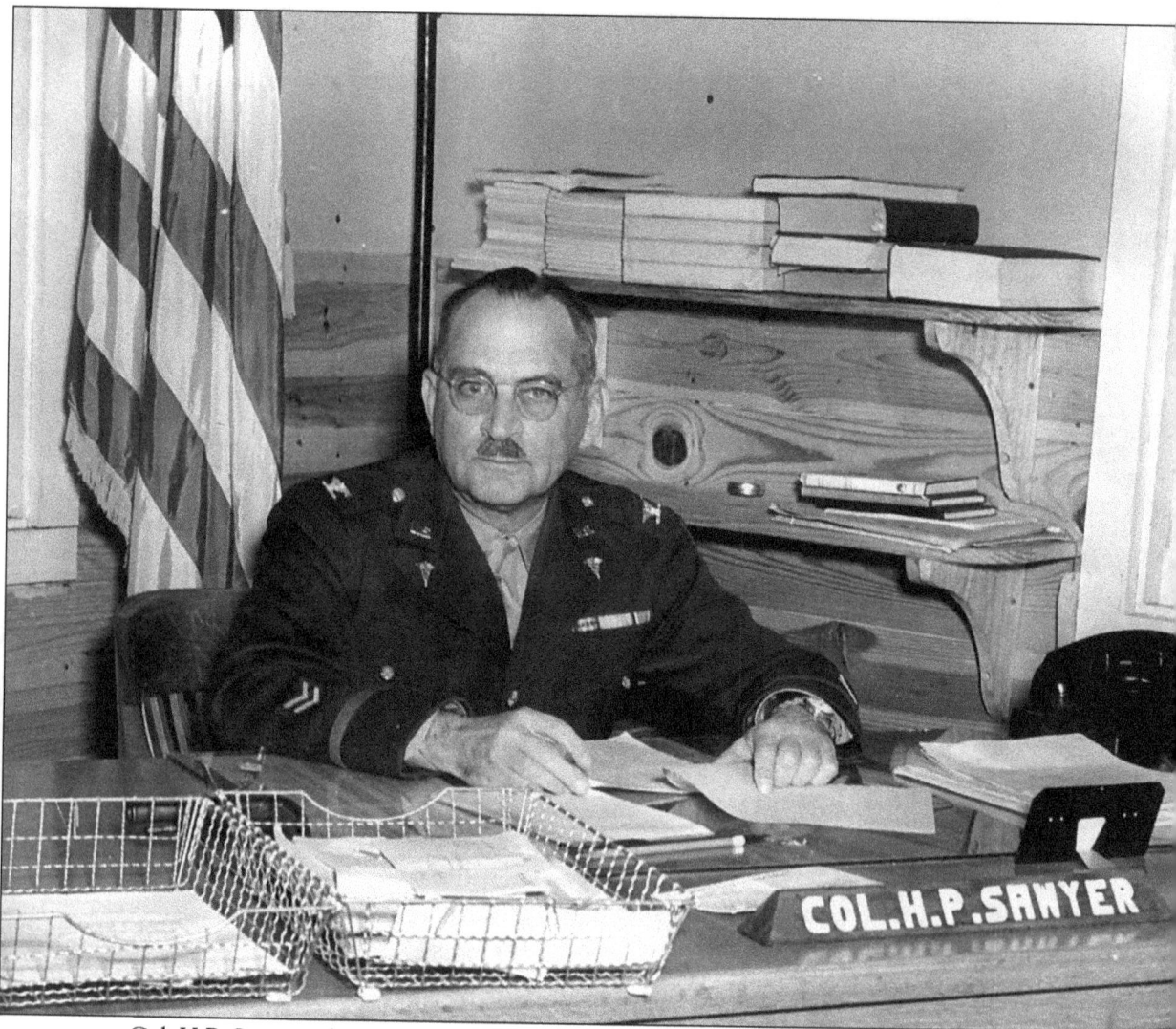

Col. H.P. Sawyer, photographed here in February of 1943, ran the camp's hospital. The hospital was a complex of 70 buildings, all connected by corridors, housing 1,375 beds. Its size made it the largest hospital in Alabama at the time. "Sick call," *Army Life* taught, "is a formation held daily to enable you to receive medical examination and treatment quickly and conveniently when you need it. . . . [Y]ou and the other men on sick call will be sent to the dispensary for examination and minor treatments. Those needing further treatment will be admitted to the hospital." (Courtesy of AAM.)

Maj. Donald B. Quisenberry served as the executive officer for the camp's Corps of Engineers detachment. Although civilian contractors were also utilized, officers such as Major Quisenberry bore significant responsibility for construction and maintenance of the camp's many facilities. (Courtesy of AAM.)

Lt. Col. Joseph Rozzell, shown here in July of 1943, was the camp's head of Ordnance and Motor Transport. By this point in the war, the tide had seemingly turned in the Allies' favor. This same month, American and British troops invaded the Italian island of Sicily, while American Marines and soldiers in the Solomon Islands continued to push back the Japanese. (Courtesy of AAM.)

Officers and soldiers working out of Camp Rucker's Range Control Headquarters coordinated the use and operation of the numerous firing ranges that dotted Camp Rucker. (Courtesy of AAM.)

Lt. Col. William A. Sullivan, shown here in April of 1943, served as the chief of surgical services at the camp's main hospital. Soldiers during World War II benefited from a host of medical advances unavailable to their fathers and grandfathers—the use of penicillin to combat infections, vaccines to prevent typhus, increased focus on treating combat fatigue and shell shock, and the use of whole blood transfusions during surgery. (Courtesy of AAM.)

In April of 1943, Capt. Raymond Fields ran the camp's Finance Branch, ensuring that the thousands of soldiers at the camp were paid timely and accurately. "Army pay must be thought of at all times in terms of the extras which go with it," rationalized *Army Life*. "Not only are America's fighting men better paid than others, but they receive the best living quarters available, good clothing and excellent food, medical and dental care, certain tax exemptions, allowances for dependents, debt relief, and free entertainment and recreation." (Courtesy of AAM.)

This photograph shows the interior of the camp's Finance Office in the spring of 1943. During World War II, what is now the Army's Finance Corps was part of the office of the Fiscal Director, Army Service Forces. In addition to pay disbursements, its responsibilities included selling war bonds and promoting National Service Life Insurance. (Courtesy of AAM.)

Col. William B. Walters, who commanded Camp Rucker's 204th Field Artillery Group, stands in this photograph in front of his adjutant in January of 1944. The group's mission was to train the field artillery units being activated at the camp. (Courtesy of AAM.)

The Ozark gate, shown here in April of 1943, stood on the eastern side of the camp. The camp's population of 30,000 quickly dwarfed that of neighboring Ozark, which only boasted 3,600 residents. (Courtesy of AAM.)

Maj. Gen. Herman F. Kramer, pictured on the left with his assistant division commander, Brigadier General Forrester, commanded the 66th Infantry Division ("Panthers"). After initial training in Florida and Arkansas, the 66th Division arrived at Camp Rucker in April of 1944 to complete its training cycle prior to overseas deployment. (Courtesy of AAM.)

Col. Arthur B. Brown, shown here in June of 1943, was the camp's dental surgeon. During World War II, the number of dental surgeons serving in the Army blossomed from 250 in 1939 to over 15,000 by the end of the war. (Courtesy of AAM.)

This photograph shows the clothing department in Post Exchange No. 2 in July of 1943. In all, there were 15 post exchanges on Camp Rucker. "The Post Exchange (PX) is your community store," *Army Life* explained, "owned jointly by you and the other enlisted men in the camp . . . The PX usually operates a general store where you may buy at very low prices all necessities and many luxuries to make life more pleasant. Patronize the PX." (Courtesy of AAM.)

Capt. Hudson P. Lipscomb stands at parade rest during a review conducted for the Second Army. Camp Rucker fell within the organizational jurisdiction of the Second Army. (Courtesy of AAM.)

Bottles line the counter at a typical camp beer bar, located at one of Camp Rucker's post exchanges. "The Army," *Army Life* soothed, "understands that although you are technically on call 24 hours a day, you need rest and recreation. It appreciates you need time to yourself; that you will be a better soldier because of this free time." (Courtesy of AAM.)

Col. Hall S. Crain Jr., as Camp Rucker's executive officer, was the installation's second-in-command. The efforts of officers like Colonel Crain would be key in the camp's effort to prepare newly activated and mobilized Army units for combat overseas. (Courtesy of AAM.)

Building 110 was constructed in 1942 as the camp's Officer's Club. Following several additions over the years, it still serves as Fort Rucker's Officer's Club today. (Courtesy of AAM.)

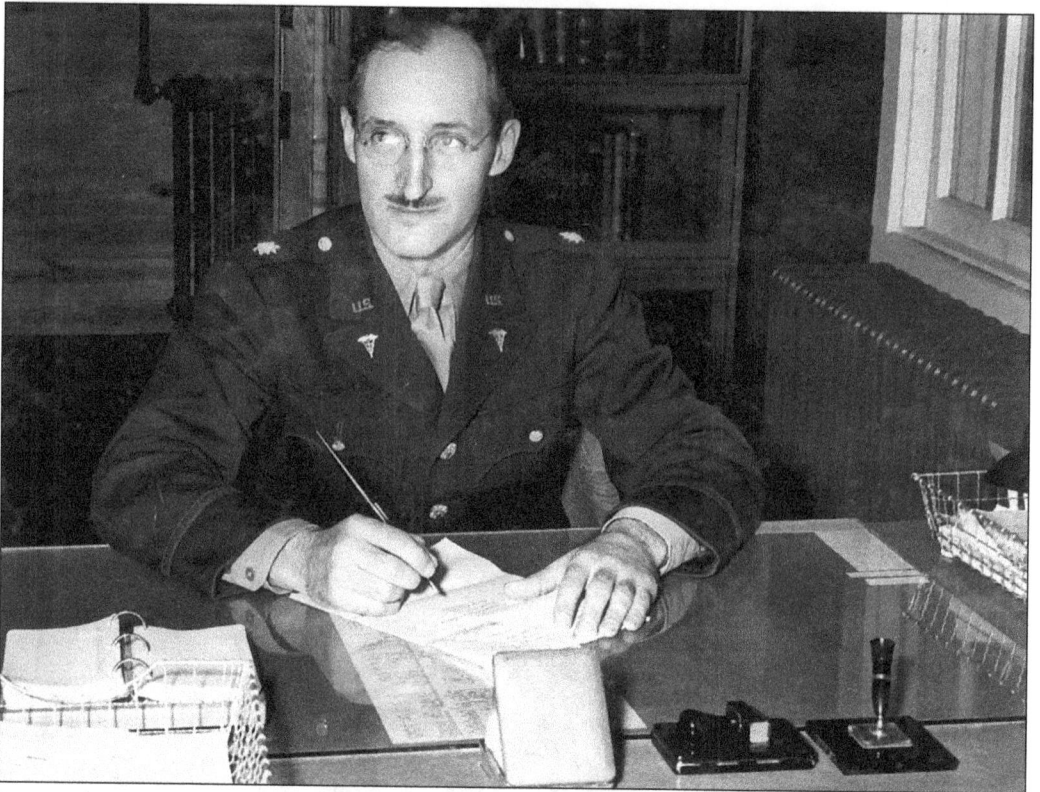

Lt. Col. Charles F. LaBelle, shown here in April of 1943, was the chief of medical service at Camp Rucker. By the end of World War II, there would be over 600,000 medical service personnel in the Army. (Courtesy of AAM.)

Col. Lloyd S. Spooner (left), commanding officer of Camp Rucker, addresses an assembly of 1,000 of the camp's civilian employees on December 8, 1943. His adjutant, Capt. Richard A. Mack, stands behind him. In 1920, as a first lieutenant, Colonel Spooner competed in the Olympic Games in Antwerp, Belgium, and collected a record total of seven medals in various shooting events. (Courtesy of AAM.)

Capt. Chang L. Hum, adjutant for the headquarters of the 18th Antiaircraft Battery and a 1938 graduate of the University of Delaware, was born in China. He is shown here in July of 1943 next to one of his unit's antiaircraft guns. (Courtesy of AAM.)

Maj. David W. Eddy, photographed in April of 1943, was the camp's billeting officer. He had the burden of ensuring the thousands of troops rotating through Camp Rucker always had adequate housing. According to *Army Life*, "A considerable party of what you can expect to learn in the Army will come from living closely with other men. The Army— the 'great equalizer'—brings you together with types of people you may never have known." (Courtesy of AAM.)

Col. A.T. Smith, the camp's executive officer, greets Alabama governor Chauncey Sparks on June 15, 1944, as he arrived for a daylong visit to the camp. As *Army Life* explained, "Yours is a people's army. In many countries the soldiers are professional, and their officers engage in political activity which often overpowers the government. In America, the Constitution specifically averted such a situation by making the Army subordinate to the elected officials of the Government." (Courtesy of AAM.)

Maj. Gen. Herman F. Kramer addresses a group of 66th Infantry Division soldiers at a unit event commemorating the division's one-year anniversary. They would deploy overseas in November of 1944. Tragically, when the division was rushed across the English Channel aboard the transport ship *Leopoldville* to reinforce American troops during the Battle of the Bulge, disaster awaited. On Christmas Eve, 1944, while soldiers sang Christmas carols on *Leopoldville's* upper decks, a German submarine planted a torpedo in the middle of the ship. When *Leopoldville* sank in the cold winter waters of the Channel, 14 officers and 748 soldiers from the division died. For the remainder of the war, the remnants of the 66th Division was used to help contain the German forces trapped in the ports of St. Nazaire and Lorient. (Courtesy of AAM.)

Two

VITAL COGS

IN THE

WAR MACHINE

"Many organizations in the Army," the War Department pamphlet *Army Life* explained, "take pride in the fact that 100 percent of their men are buying War Bonds. These purchases are important for two reasons. First, more than 90 cents of every dollar you put into bonds is used directly to help pay for the war. Second, after you have been discharged and return to your home, your War Bonds can provide you with ready cash to tide you over while you get reestablished in civilian life." This photograph, taken in January 1944, shows Maj. R.P. Fields receiving reports on the camp's War Bond drive from Edward C. Kearny, of the Post Engineers, and Marilyn McNally, of the camp's Supply Division. (Courtesy of AAM.)

"The Army is a pretty human," *Army Life* proclaimed. "It knows that letters will mean a great deal to you. It not only provides many places around camps where they may be written; it delivers your mail as promptly as possible." Here, mail call takes place at the 311th Ordnance Company. (Courtesy of AAM.)

Loads of incoming mail for lonely soldiers far from home threatened to swamp Camp Rucker's postal facilities during the 1943 Christmas season. On the left, postal clerk Dewey L. Brannon of Hartford, Alabama, instructs Pvt. Leonard Hatcher of Raleigh, North Carolina, regarding proper mail sorting. Hatcher, along with James Gulfto of Patterson, New Jersey (far right), and Cpl. Pete Lanzo of Chelsea, Massachusetts (second from right) were assisting the regular civilian mail clerks during the Christmas rush. In the middle of the group stands postal clerk John D. Lassister, also of Hartford, Alabama. (Courtesy of AAM.)

Mail orderlies assist with the 1942 Christmas rush. "One of your privileges as a soldier," explained *Army Life*, "is mailing without postage your personal letters, post cards, and V-Mail. Include your name, rank, serial number and organization in the return address in the upper left-hand corner of your envelope or post card; the word "Free" must appear in the upper right-hand corner in your own handwriting." (Courtesy of AAM.)

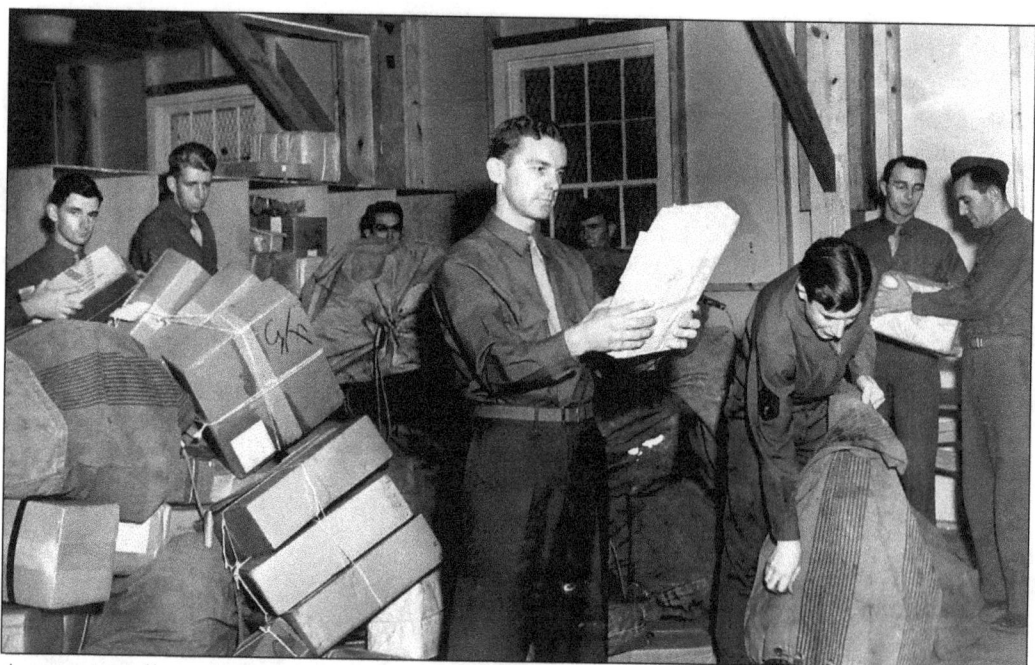

A private pulling duty as a mail orderly reads the address on a package arriving for one of his fellow soldiers during the Christmas season of 1942. Elsewhere in the war that Christmas, American troops battled Germans in Tunisia and Japanese in the Solomon Islands, while the Germans and Soviets struggled for control of Stalingrad. (Courtesy of AAM.)

Mrs. Mittie Ward (immediate front) and Mrs. Missouri Russell (behind Mrs. Ward) worked in the camp's salvage sewing operation. They, and scores of women like them, would work to repair damaged or worn uniforms and other items of equipment. Reportedly, the camp's salvage operation saved the Army a remarkable $750,000 a month. (Courtesy of AAM.)

A worker in the camp's Quartermaster Division, Automotive Branch, stencils an address on a carton of motors bound for Atlanta in April of 1943. That same month overseas, the United States extracted a measure of revenge for Pearl Harbor by shooting down the plane carrying Admiral Yamamoto, Japan's architect of the Pearl Harbor attack. (Courtesy of AAM.)

Office clerks in the automotive branch of the camp's quartermaster division appear in the photo above. Notice the specially constructed "perpetual inventory desks" containing cards representing all items in Camp Rucker's automotive parts warehouse. (Courtesy of AAM.)

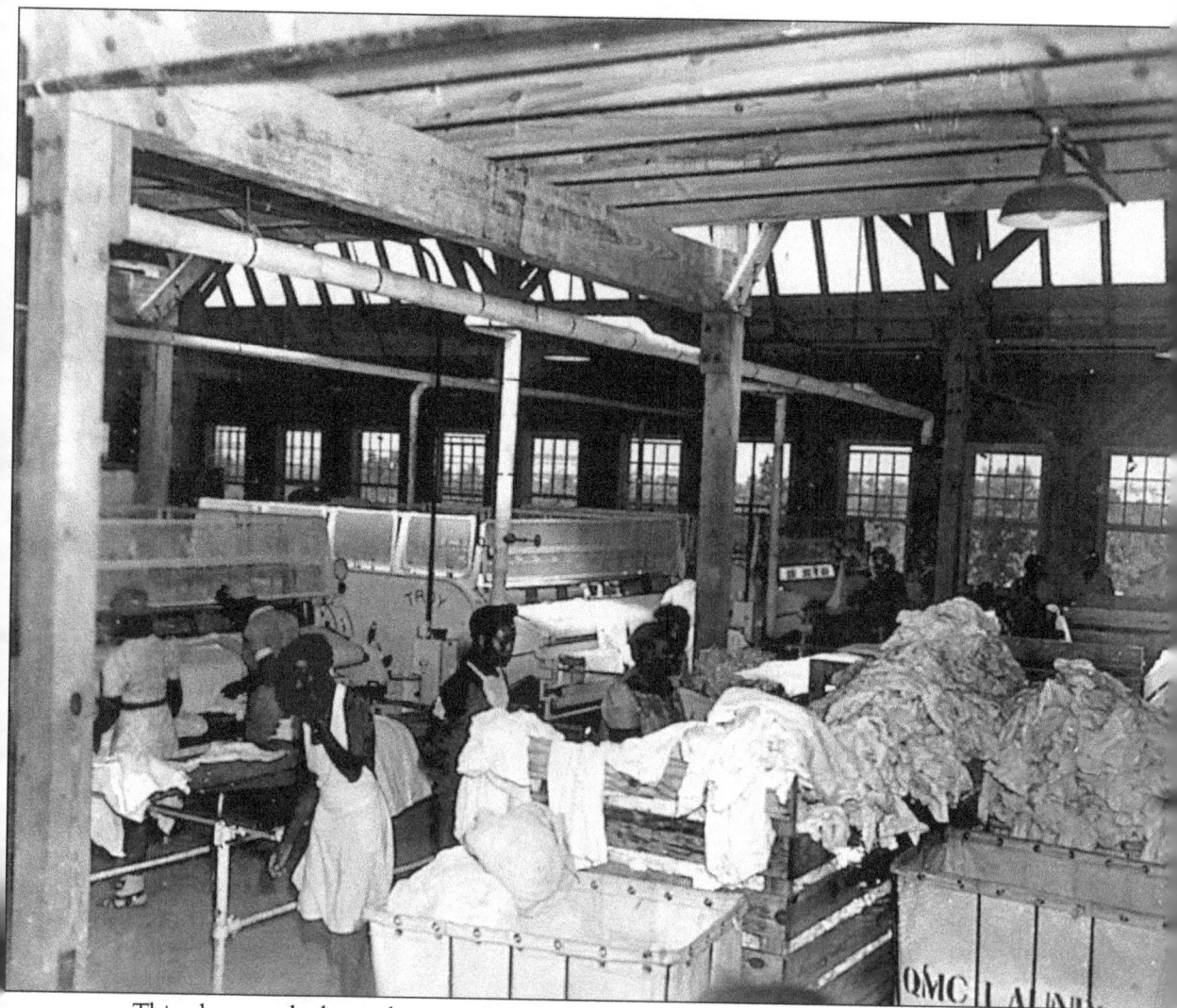

This photograph shows the camp laundry in the summer of 1942. "In order to have frequent changes of clothing," *Army Life* warned, "you may wash some of your laundry yourself. If you do, be careful with woolens. Hot water will cause them to shrink. Stretch them back to size after washing. Do not use soaps containing lye on clothes whose color you want to keep." (Courtesy of AAM.)

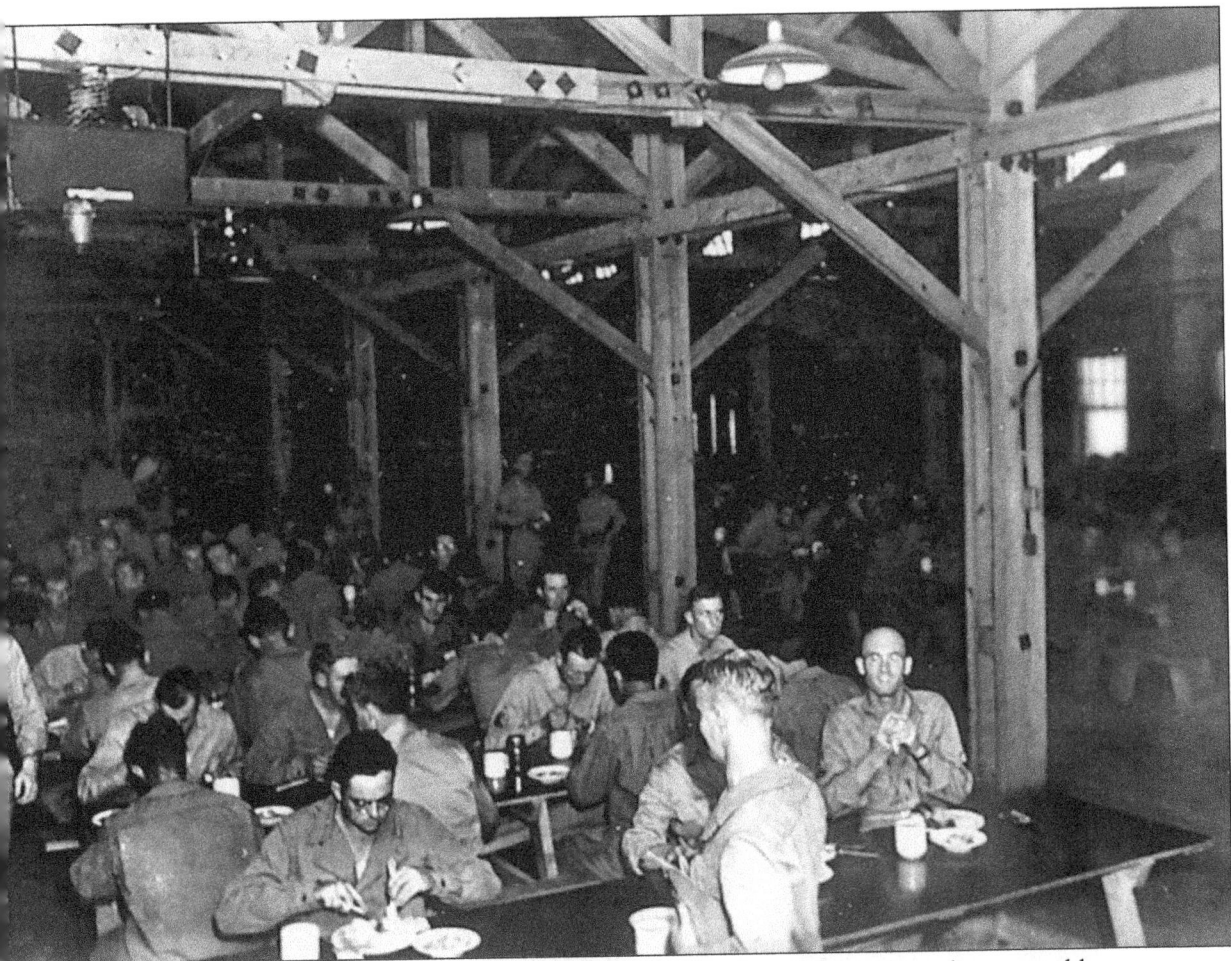

Camp Rucker boasted 14 mess halls such as the one shown in this photograph, each one capable of accommodating 500 soldiers at a time. "Within their first six months in the Army," *Army Life* stated, "most inductees gain an average of six pounds per man. They eat three full meals a day, at regular hours. They learn to eat some foods they never saw before, and to like some dishes they never liked before." (Courtesy of AAM.)

The soldiers pictured in this March 1943 photograph are checking and sorting shoes turned in by their fellow soldiers from salvage. From left to right are Pfc. Joseph L. Bilodeau of Manchester, New Hampshire; Pvt. George A. Kessaris of Danvers, Massachusetts; and Cpl. Harry Pendroff of Brooklyn, New York. (Courtesy of AAM.)

Lieutenant Huss, the salvage officer for the camp's Quartermaster Office, checks in garments returned for salvage. "If you gain or lose weight so that in time your clothes no longer fit," *Army Life* explained, "you may exchange the garments at your supply room for others of proper size. If the clothing becomes unusable through fair wear and tear, you may turn it in as salvage." (Courtesy of AAM.)

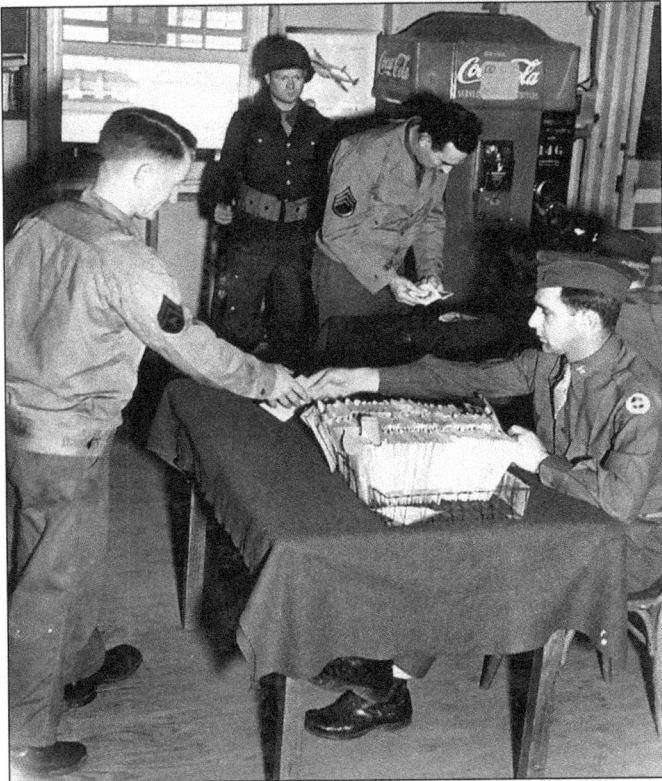

A guard stands watch during payday at Camp Rucker during the winter of 1942. Privates earned $42 a month, which was later raised to $50 a month by 1944. The second lieutenants leading their platoons earned $150 a month. Winning the Medal of Honor qualified its recipient for an extra $2 a month. (Courtesy of AAM.)

By the late summer of 1944, the absence of so many men away in the service left the farming communities around Camp Rucker with a significant manpower shortage. That shortfall factored in the War Department's decision to send approximately 2,000 German and Italian prisoners of war to Camp Rucker. Here, a group of Italians pose for a photograph in one of the camp's kitchens. (Photo by U.S. Army Signal Corps, courtesy of AAM.)

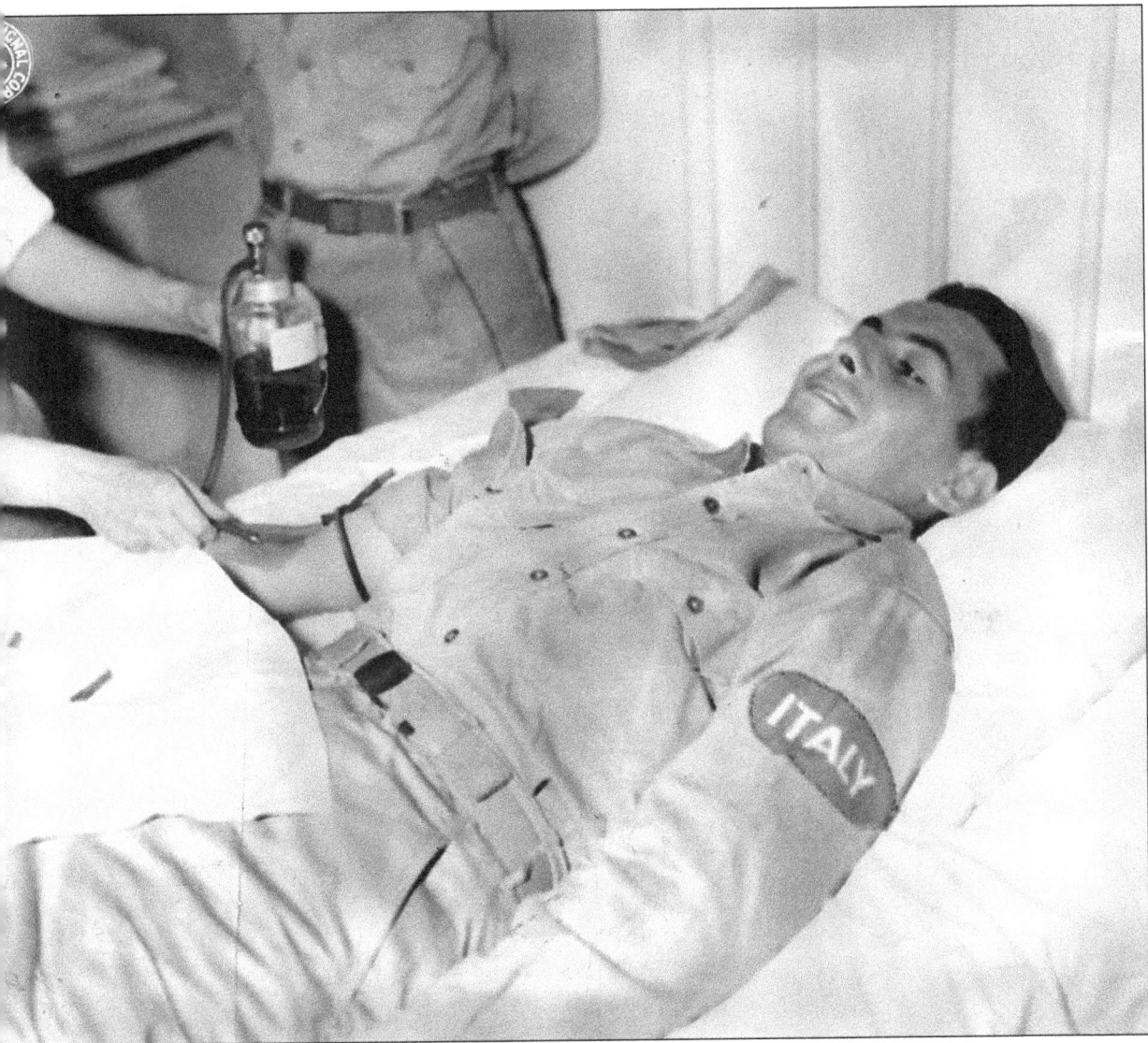

The Italian prisoners existed in somewhat of a nether zone in the eyes of the United States military. Although they had been captured in battle in such places as North Africa, Sicily, and southern Italy, their government had surrendered in July of 1943 to the Allies. In response, Nazi Germany occupied the northern part of their country and formed a puppet state, the Italian Socialist Republic. Because of this unusual situation, the Army paroled its Italian prisoners into so-called Italian Service Units. In this photograph, an Italian soldier donates blood at a hospital in nearby Dothan. (Courtesy of AAM.)

Members of the Italian Service Unit attend English class. These men, used on labor details at Camp Rucker and in surrounding communities, were paid $24 a month. (Photo by U.S. Army Signal Corps, courtesy of AAM.)

Members of the camp's Italian Service Unit pose in front of their unit's canteen, where they could buy Cokes, beers, and cigarettes. (Photo by U.S. Army Signal Corps, courtesy of AAM.)

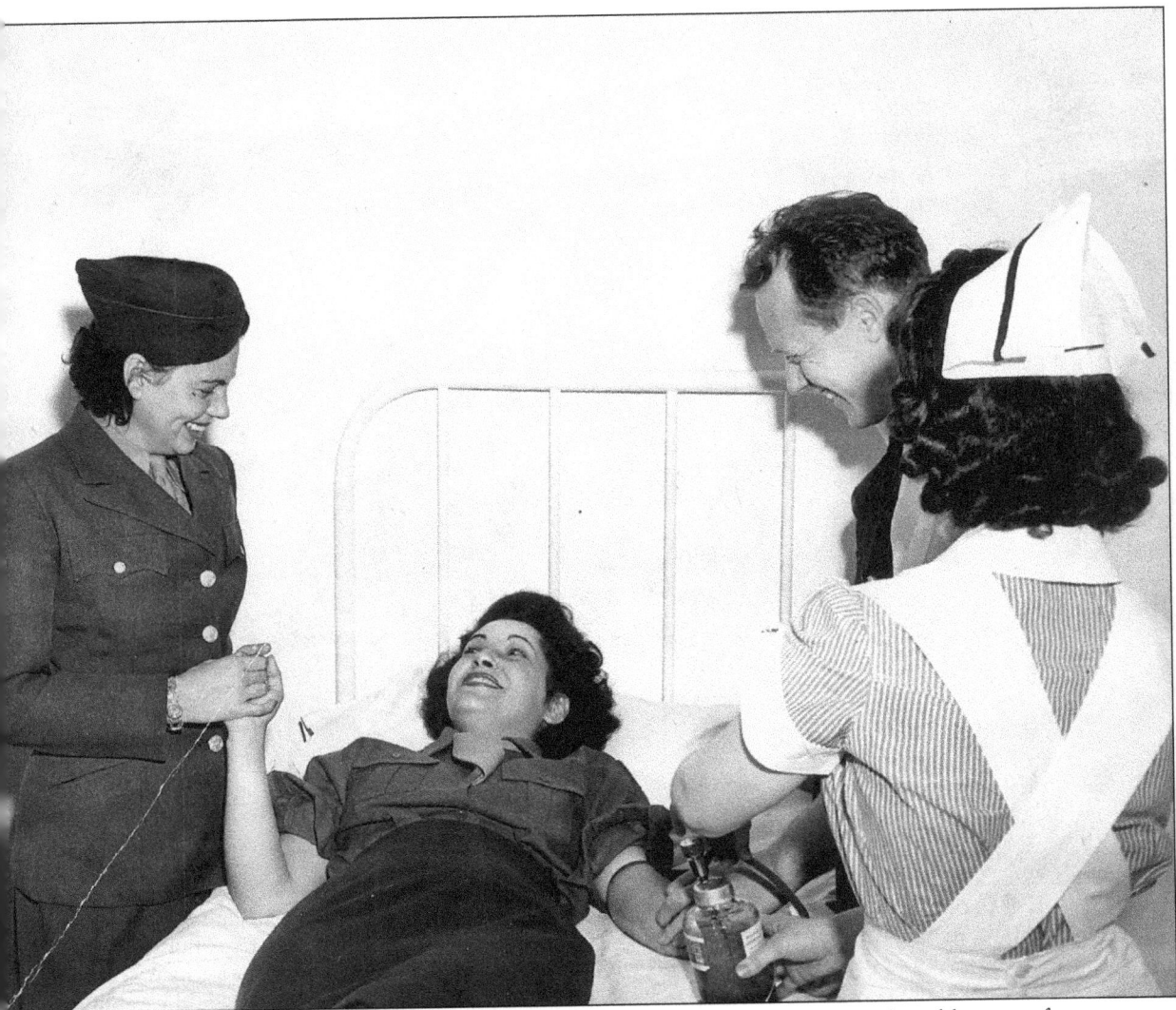

In this photograph, a member of the Women's Army Corps, or WAC, is comforted by one of her comrades as she donates blood. In May of 1942, the Army had created, on a trial basis, the Women's Army Auxiliary Corps in an effort to address the manpower shortages it faced. The WAAC's success led to the full-scale integration of women into the Army on July 1, 1943, with the creation of the Women's Army Corps. By 1945, there were more than 100,000 WACs in uniform. (Courtesy of AAM.)

This photograph shows the convening of a court-martial at Camp Rucker in January of 1944. Article 96 of the Articles of War, *Army Life* warned its readers, covered a variety of unspecified offenses: "disorderly conduct and bringing discredit upon the military service, by such acts as not paying debts, writing checks you cannot cover, lending money for interest, and gambling in violation of orders. Even such minor acts as thumbing rides or wearing a dirty uniform can be punishable under this Article. The court-martial will determine what sentence to impose for these acts of misconduct. Remember one thing: you can't 'get away' with anything. The Army has been in business for a long time and is always one step ahead of you, so don't try to outsmart it." (Photo by U.S. Army Signal Corps, courtesy of AAM.)

Col. Lloyd S. Spooner pins the Emblem for Civilian Service on George E. Walter, Station Complement Area Engineer Division, recognizing a minimum of six months consecutive service at Camp Rucker. Colonel Spooner's adjutant, Capt. Richard Mack, stands behind him. Also receiving awards were (left to right) Joe Jernigan Jr., Lillian Kirch, Francis Adams, Ann Dolvin, and Amy K. Johnson. (Photo by U.S. Army Signal Corps, courtesy of AAM.)

This photograph, taken in March of 1943, shows an unusual subject—a group of men identified only as "Chinese military personnel on duty at [the] junior officer's mess." (Courtesy of AAM.)

From left to right, Cpl. Harry Pendroff, Cpl. Delmas Aaron, and Pvt. George A. Kessaris sort through equipment turned in by the camp's soldiers for salvage during March of 1943. (Courtesy of AAM.)

WAC Pvt. Frances Knueven operates a comptometer at Camp Rucker's Finance Office in August of 1943. "Members of the Women's Army Corps are your sisters-in-service," *Army Life* preached. "If you think that the Army life is too tough for a girl—for your own sister, for example—you must have even greater respect for the girls who are going through with this tough job." (Photo by U.S. Army Signal Corps, courtesy of AAM.)

Three

THE FIELDS
OF FRIENDLY STRIFE

"Upon the fields of friendly strife," Gen. Douglas MacArthur once said, "are sown the seeds that, on other days and on other fields, will bear the fruits of victory." Shown here are the ballplayers of Company B's baseball team in the 84th Chemical Mortar Battalion. These soldiers' battalion would later land in Salerno, Italy, in 1944, and help the U.S. Army capture Rome and battle its way up the boot of Italy. (Courtesy of AAM.)

The men of the 13th Chemical Maintenance Company's bowling team pose for the camera in the winter of 1942. "Your organization fund and your officers will provide a great variety of organized and individual sports," *Army Life* stated. "You may want to let off steam with a basketball or baseball game, or you may want nothing more strenuous than a game of pool or table tennis, but in either cases the competition and the fellowship of playing will give you a lot of fun . . . participate!" (Courtesy of AAM.)

The Camp Rucker station hospital's nurses fielded a basketball team in the winter of 1942. Camp Rucker served as a basic training center for nurses entering military service following the completion of their training. Eventually, approximately 2,080 Army nurses graduated from the camp's eight-week nurses basic training course. (Courtesy of AAM.)

The Camp Rucker station hospital's nurses are captured in an impromptu group photograph in the winter of 1942. (Courtesy of AAM.)

Camp Rucker boasted a well-equipped varsity basketball team in the fall of 1942. In November, as these soldiers played basketball at Camp Rucker, their brothers in arms were landing on the

beaches of North Africa during Operation Torch. Within a matter of weeks and months, these men would likely follow in their footsteps. (Courtesy of AAM.)

These five men formed the basketball team for Company A, 336th Engineer Combat Battalion. (Courtesy of AAM.)

This photograph shows the Special Hospital Unit's basketball team during the winter of 1943. The officer on the left is Major Cohen. (Courtesy of AAM.)

Lieutenant Distasio of the 336th Engineer Combat Battalion helps tape up Private Wood prior to a boxing match in the summer of 1942. In June of that same summer, the American Navy handed a decisive defeat to the Japanese at the Battle of Midway. (Courtesy of AAM.)

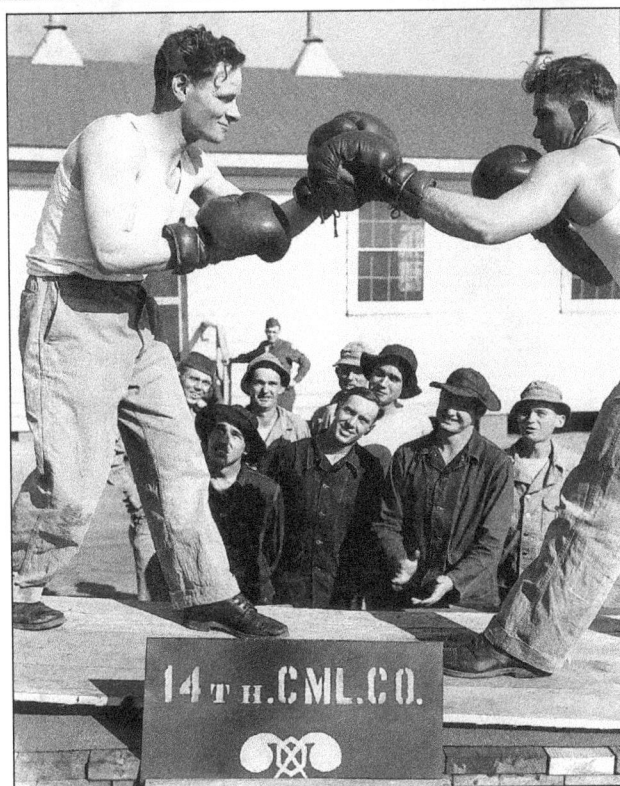

In this photograph, two soldiers of the 14th Chemical Maintenance Company square off in a boxing match in the summer of 1942. The 14th was formally activated at Camp Rucker in August of 1942 and later landed in Europe on June 26, 1944. (Courtesy of AAM.)

The Special Hospital Unit's basketball team takes on the team fielded by the 209th Ordnance Battalion in the winter of 1943. (Courtesy of AAM.)

This photograph shows more action between the men of the Special Hospital Unit and the 209th Ordnance Battalion on the basketball court. (Courtesy of AAM.)

The camp's varsity basketball team plots out a play in February of 1943. That same month, the besieged Germans surrendered to the Soviet Red Army at Stalingrad. (Courtesy of AAM.)

The men of the 14th Chemical Maintenance Company enjoy a rough game of "Red Rover" during the fall of 1942. The mission of a chemical maintenance company was to repair chemical warfare weapons and protective equipment and to assist in other salvage work and decontamination as necessary. (Courtesy of AAM.)

A line of 14th Chemical Maintenance Company soldiers staggers before the surge of an onrushing "Red Rover" player. "The benefits of physical exercise under conditions of reasonable living habits are self-evident," *Army Life* lectured. "You find that you can not only do more work, but you enjoy doing it. This enjoyment results from the lack of fatigue, and this in turn results from the large reserve of energy which exceeds the demands of the work." (Courtesy of AAM.)

Even the camp's female civil service employees fielded a basketball team in the fall of 1942. (Courtesy of AAM.)

The dirt patch between a pair of barracks served as a volleyball court for the soldiers of the 14th Chemical Maintenance Company. (Courtesy of AAM.)

Sergeant Dlugiewicz was hailed as one of the camp's "bowling experts" in February of 1943. (Courtesy of AAM.)

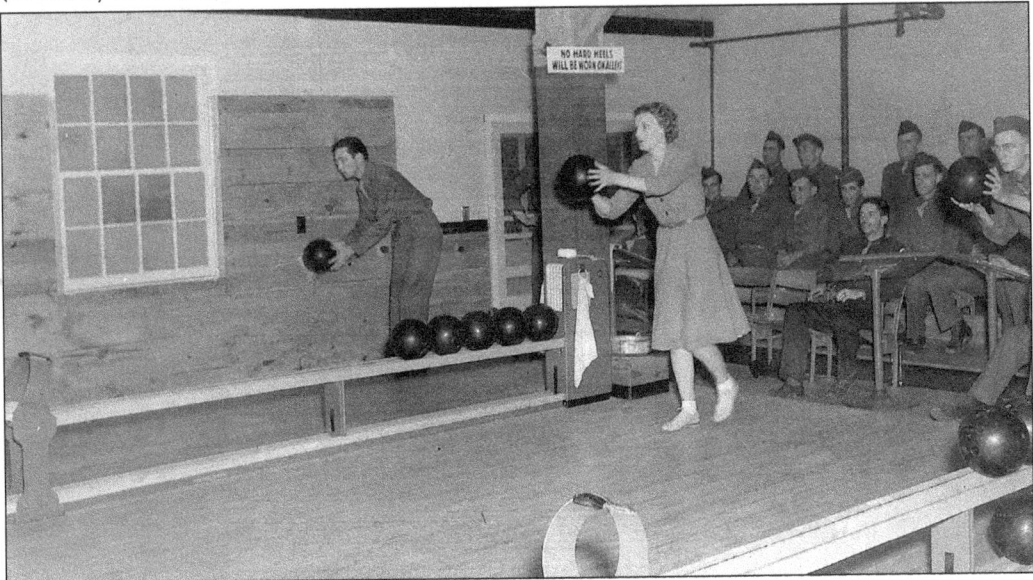

Post Exchange No. 3 boasted a bowling alley, shown here in the winter of 1942–1943. (Courtesy of AAM.)

Tony Kareb, on the left, takes on Nick Serio in a wrestling match in the camp's field house. (Courtesy of AAM.)

Major Cohen stands proudly beside the men of the Special Hospital Unit's basketball team in the winter of 1943. (Courtesy of AAM.)

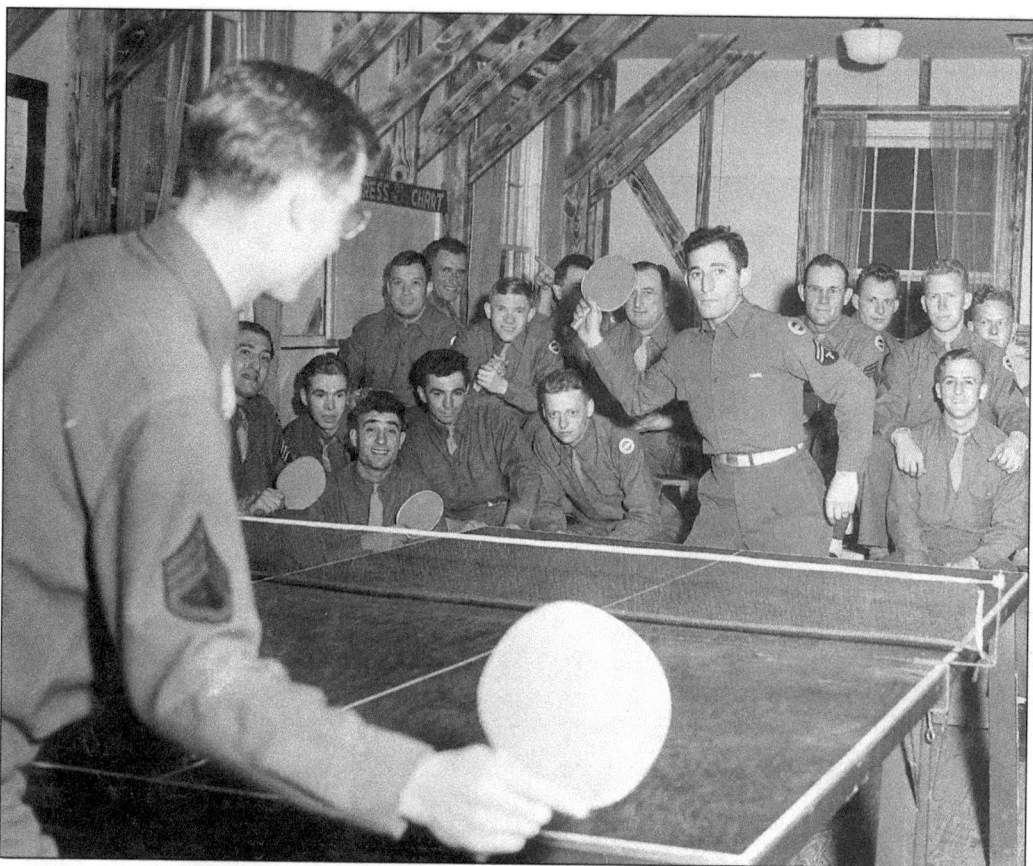

A group of soldiers gather for a ping-pong tournament in the winter of 1942. (Courtesy of AAM.)

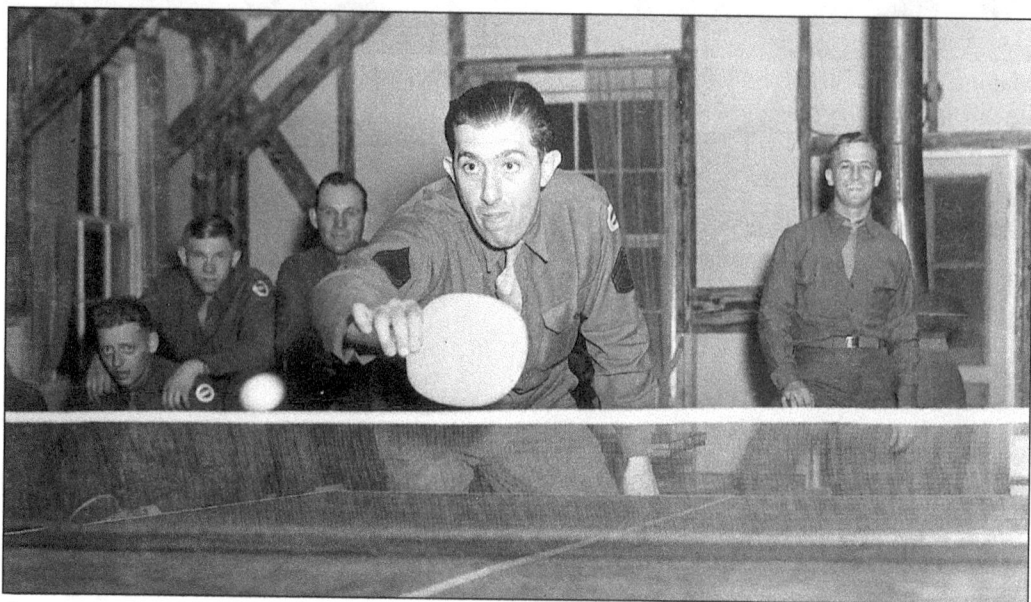

A sergeant shows his form during a spirited game of ping-pong. (Courtesy of AAM.)

Four

REST AND
RELAXATION

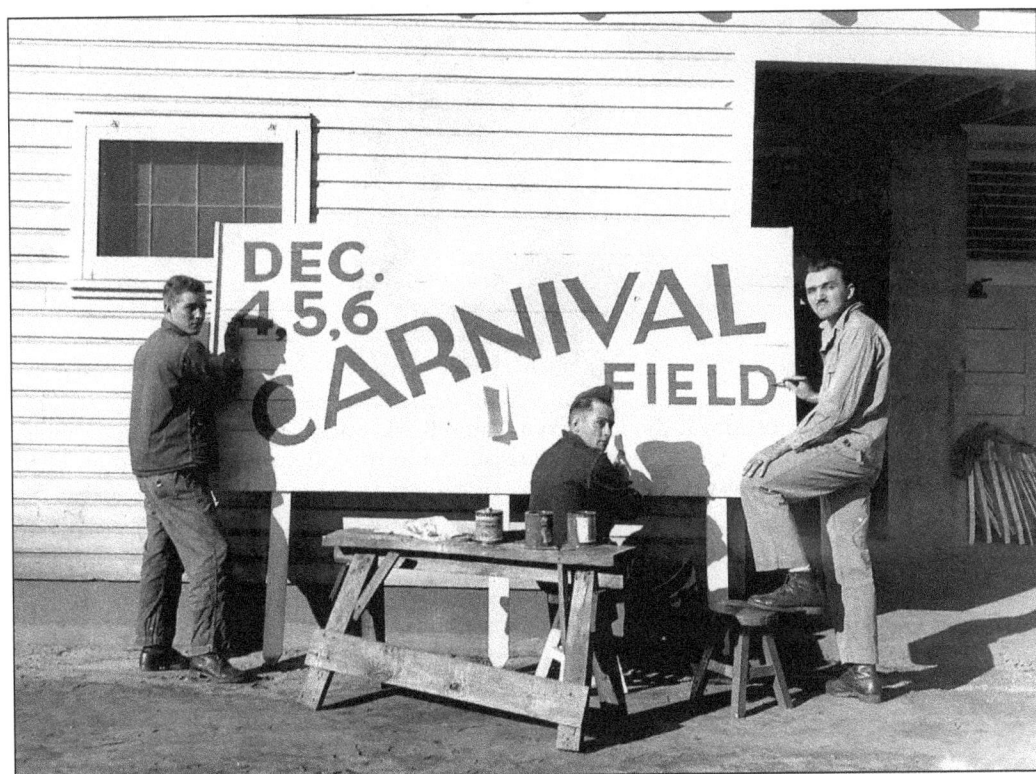

In December of 1942, Camp Rucker celebrated with a winter carnival. Here, a group of men paint a sign in preparation of the event. Don't use your free hours, advised *Army Life*, "just holding down a bunk in the barracks; don't seclude yourself from the other men; don't be timid about participating in activities or enjoying the facilities available. Simply *doing something* is good for your state of mind." (Courtesy of AAM.)

Dothan, Alabama, was just down the road from Camp Rucker and was home to Napier Field, an Army Air Corps training base. The three girls shown in this photograph are Dothan High School cheerleaders. (Courtesy of AAM.)

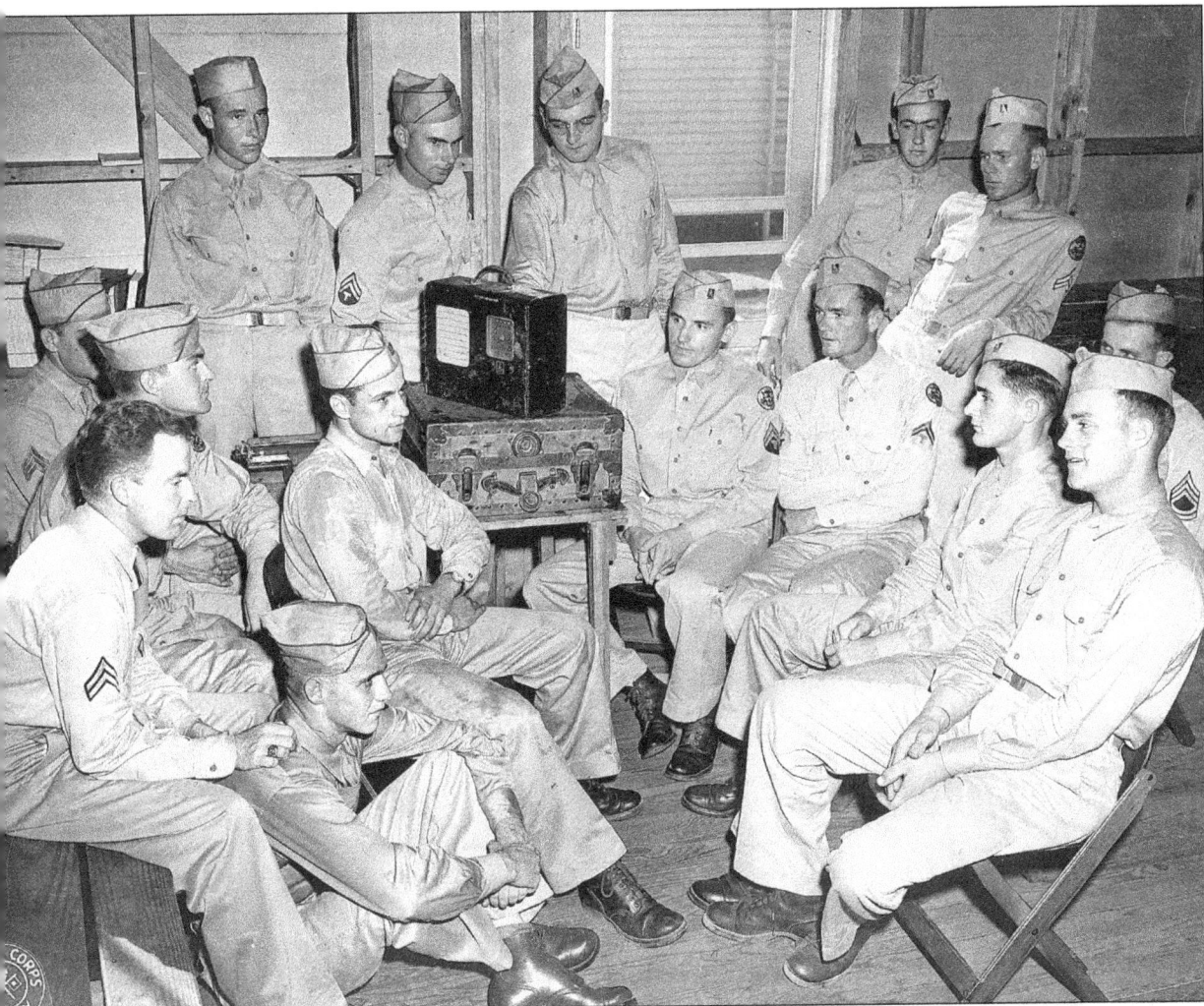

Members of the 630th Tank Destroyer Battalion listen to a radio program produced by the camp's Public Relations Office describing their unit and training. These soldiers would eventually see vicious fighting in Europe trying to stem the initial German attacks through the Ardennes Forest during the Battle of the Bulge. (Photo by U.S. Army Signal Corps, courtesy of AAM).

Jerome Currie, seated at the head of the table, was an Associated Press reporter who accompanied a visit by Alabama's governor on his trip to Camp Rucker on November 10, 1944. (Courtesy of AAM.)

Col. Lloyd S. Spooner pays M.Sgt. Troy Miller, the president of the camp's NCO Club, a dollar for the first ticket to the club president's birthday benefit dance. M.Sgt. George T. Byrd sits on the left. (Photo by U.S. Army Signal Corps, courtesy of AAM.)

The soldiers of the 8th Tank Group helped commemorate the Christmas season by building this life-size Nativity scene in front of Hilltop Chapel No. 9. The 8th Tank Group, with soldiers scattered among several Army camps, was the unit responsible for training the tank units being activated at Camp Rucker. The group would eventually form the cadre for the 8th Armored Division. (Courtesy of AAM.)

Soldiers gather in the library of Service Club No. One during the summer of 1942. Service clubs such as this one were an important part of maintaining the soldiers' morale but, as *Army Life* lectured, there was more to morale than what such clubs could provide. Good morale "is knowing that you are in an Army that has never lost a war. It is belonging to the company with the squarest commanding officer who ever lived; to the platoon with the drill award to its credit; to the squad with the greatest guys in the world." (Courtesy of AAM.)

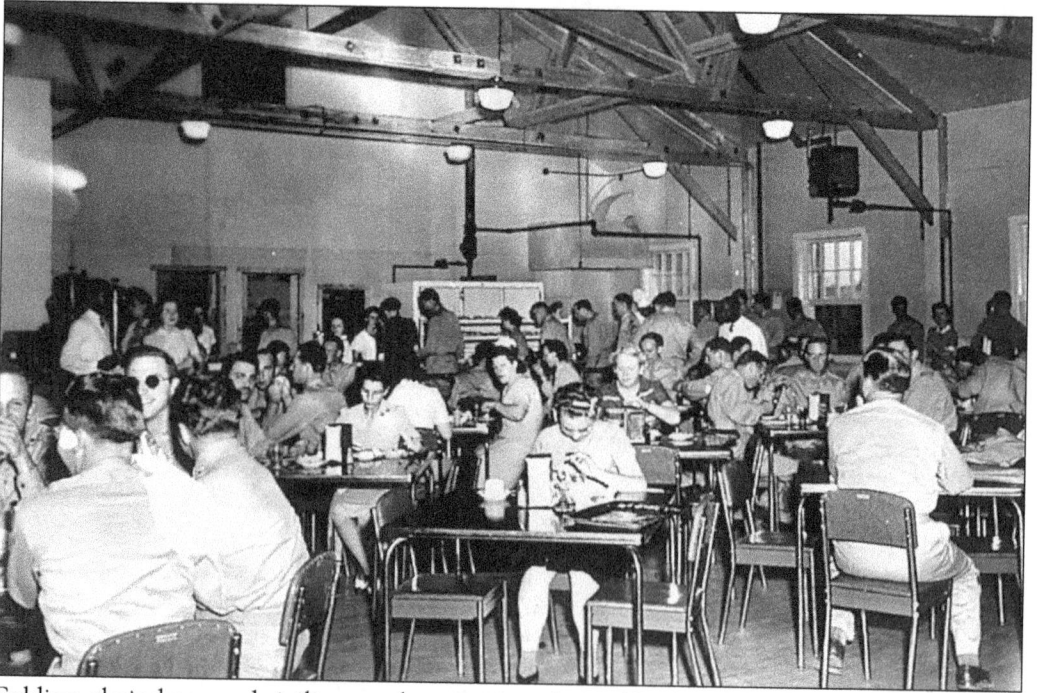

Soldiers, their dates, and civilian employees enjoy a bite of food in the cafeteria of Service Club No. One during the summer of 1942. (Courtesy of AAM.)

A group of soldiers, taking a break from the noise and tumult of barracks life, read a collection of books and magazines in one of the camp's day rooms. "Your organization probably has a day room, furnished by money from your organization fund," *Army Life* explained. "This is usually less formal and more masculine than the service club. There are no hostesses, no guests. There is no planned program. You may swap yarns, listen to the radio, write letters, read, or play games." (Courtesy of AAM.)

In the fall of 1942, these officers and soldiers had the pleasant task of selecting Miss Camp Rucker. In the middle sits Captain Schiavelli. On his left sits Lieutenant Owen. (Courtesy of AAM.)

In January of 1943, this young lady was crowned Miss Camp Rucker. (Courtesy of AAM.)

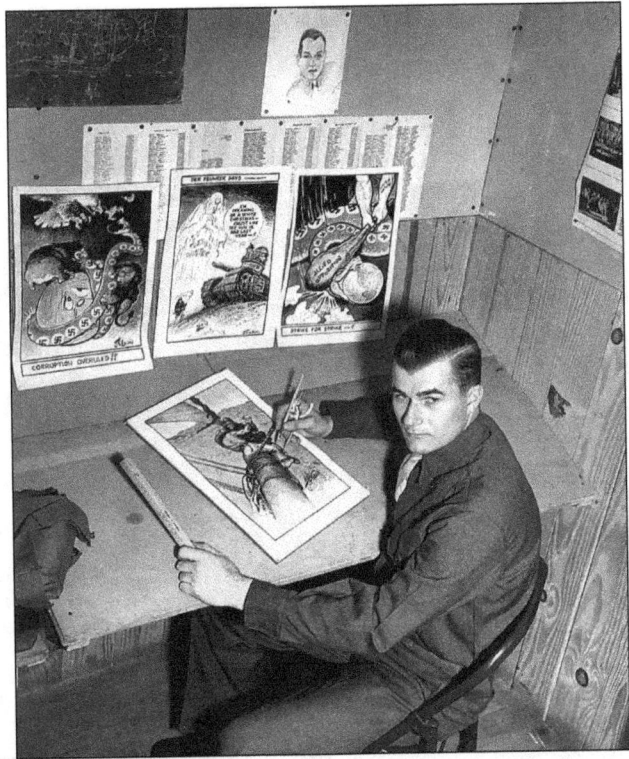

Pvt. J. Olsen, of the 37th Medical Ambulance Battalion, sits at his drawing table. Several of his posters line the wall in front of him. (Courtesy of AAM.)

Pvt. Guy Wood, of the 84th Chemical Mortar Battalion, sits at the keyboard and entertains the troops with one of his songs. Smoke, high explosives, and white phosphorus shells launched from the 84th's mortars would eventually assist the grueling American advance up the boot of Italy. (Courtesy of AAM.)

A happy crowd of soldiers gathered in Post Theater No. 3 to watch a performance of the play *Claudia* in the fall of 1942. (Courtesy of AAM.)

This soldier and his guest enjoyed front row seats at *Claudia* in 1942. In addition to the occasional play, Camp Rucker offered movies in the after-duty hours. "The United States Army Motion Picture Service operates one or more theaters in your camp. Admission is 15 cents . . . . Often your theater will play the first-run Hollywood pictures weeks and even months before they are shown in town." (Courtesy of AAM.)

This photograph shows an overhead view of the action at the 11th Hospital Center's dance at Service Club No. 1. (Courtesy of AAM.)

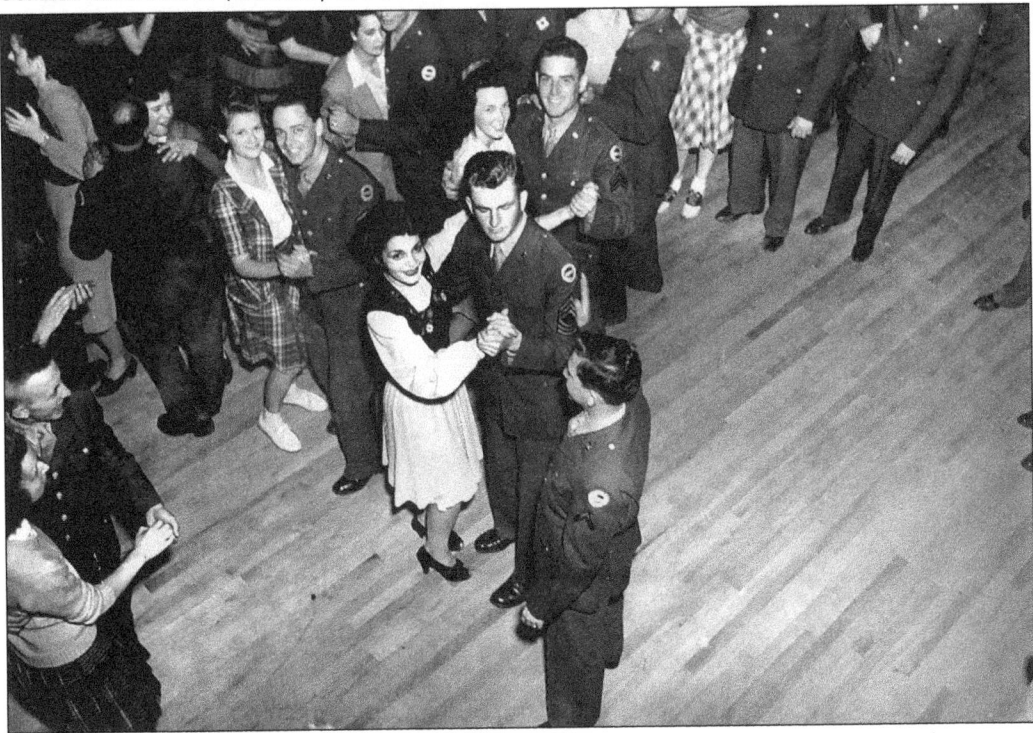

One can only wonder what bold statement this buck private is making in front of the master sergeant and his companion at the 11th Hospital Center's dance. (Courtesy of AAM.)

Maj. A.A. Hall conducted the dance orchestra of the 9th Hospital Center during this dance in the fall of 1942. (Courtesy of AAM.)

The featured entertainment at the camp's 1942 United Nations Christmas Party was the Red Army's visiting Gen. Platoff Don Cossack chorus. (Courtesy of AAM.)

The Don Cossack chorus also performed a Christmas show in 1942 for the patients and staff of Camp Rucker's station hospital. (Courtesy of AAM.)

The so-called Grand March marked the beginning of Camp Rucker's first Ward Bonds dance. Here, in the front rank, Brigadier General Manley walks with Miss Cooper on his right arm and his wife on his left. Colonel Ekedahl is on Mrs. Manley's left arm. (Courtesy of AAM.)

These musically inclined soldiers formed the 11th Hospital Unit's "jive combo." Warrant Officer Campbell, standing at the right front of the group, led the combo. (Courtesy of AAM.)

An amused USO entertainer visiting Camp Rucker in March of 1943 shows off her form on the 336th Engineer Combat Battalion's obstacle course. (Courtesy of AAM.)

Lieutenant Yeoman emceed Camp Rucker's first War Bonds dance in the winter of 1942. (Courtesy of AAM.)

When the 23rd Chemical Company threw a dance, the male-to-female ratio, like most dances at Camp Rucker, left quite a bit to be desired—at least as far as the soldiers were concerned. (Courtesy of AAM.)

The cooks of the 48th Hospital Unit dazzled their fellow soldiers with their dance moves at Camp Rucker's winter carnival in 1942. (Courtesy of AAM.)

The dancing cooks shown in the preceding photograph had some serious competition from this entertainer at the winter carnival. (Courtesy of AAM.)

These soldiers are pitching pennies at a booth at the camp's 1942 winter carnival. (Courtesy of AAM.)

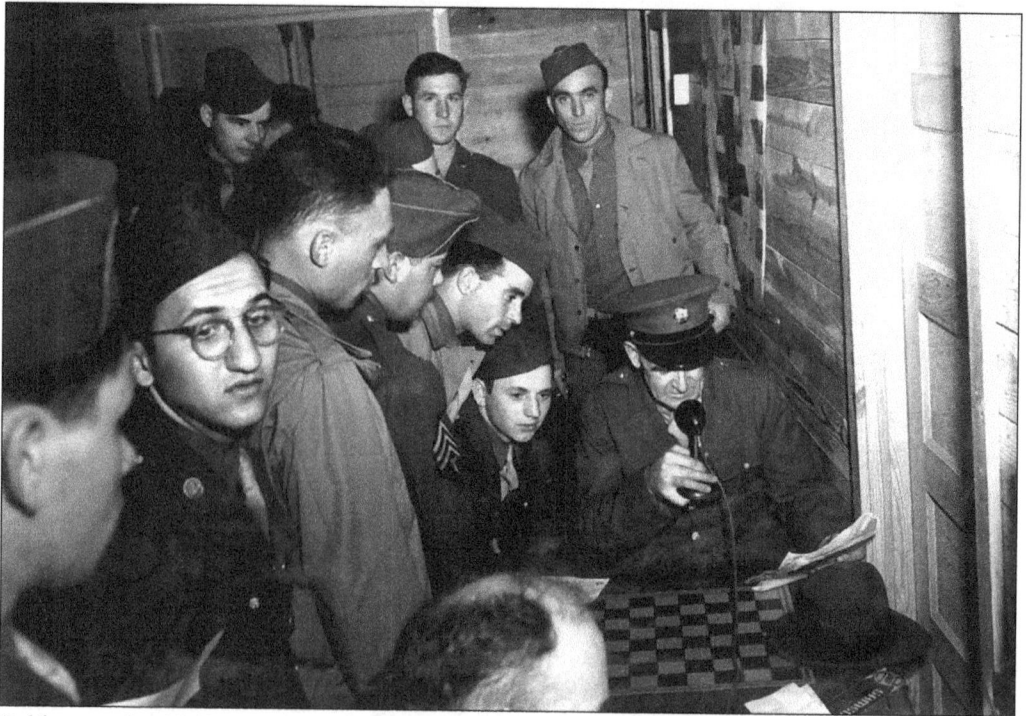

Soldiers at the 1942 winter carnival had a chance to record phonograph records to send home to family and friends. (Courtesy of AAM.)

This dancer was another popular attraction at Camp Rucker's 1942 winter carnival. (Courtesy of AAM.)

In February of 1943, the Bobby Byrne Orchestra broadcasted nationwide from Camp Rucker. Coca-Cola sponsored the event. In 1943, Bobby Byrne would have shared the radio waves with the likes of Frank Sinatra, Tommy Dorsey, and Glenn Miller. (Courtesy of AAM.)

A chaplain conducts a church service before a congregation of soldiers and WACs. "Religion helps you face hard duty," *Army Life* advised. "Every camp in the country and every organization in action overseas has facilities for divine worship. The chaplains who serve in the camps and with the fighting men extend their efforts and compassion into every element of their men's lives." (Photo by U.S. Army Signal Corps, courtesy of AAM.)

The Camp Rucker Dance Orchestra's first gig was the farewell party for General Manley on August 31, 1943. From left to right, the orchestra members are Pfc. George Haggard, Pfc. Roman Kamorski, Pfc. Udall Mason, Pfc. Chad Coyle, Pvt. Arthur Horvah, Pfc. Kay DeMarch, Sgt. Samuel Orto, Cpl. Arthur Beall, and orchestra director Pfc. Thomas F. Caruso. (Photo by U.S. Army Signal Corps, courtesy of AAM.)

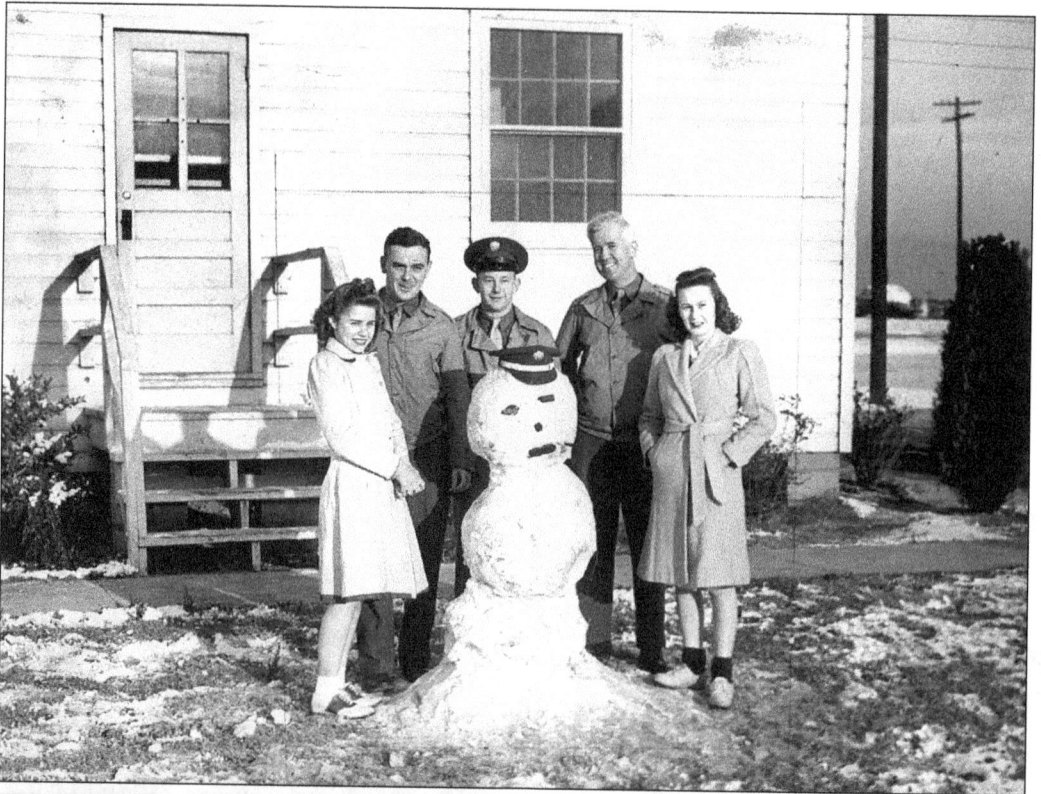

A rare snow came to Camp Rucker in December of 1943, which enabled the construction of a short-lived snowman in front of the camp's Provost Marshal's building. From left to right stand Maureen Hughes of Dothan, Cpl. William Martin, Pvt. Robert Hoenike, Pvt. Joe Fearon, and Mrs. Claire Cannon of Enterprise. (Photo by U.S. Army Signal Corps, courtesy of AAM.)

Lucille Schmidt, a medical stenographer at Detroit Edison Company in Detroit, Michigan, made the long trip south to Camp Rucker in September 1944 to visit her boyfriend Ray Roberts. They would later marry in 1946. Lucille is sitting in front of one of the camp's service clubs in this photograph. (Courtesy of Ray Roberts.)

Pictured from left to right are Ray Roberts, Lucille Schmidt, and Tommy Mitchell. Roberts was an automobile mechanic and Mitchell was a company clerk with the 766th Ordnance Company, a unit of the ill-fated 66th Division. The 66th lost 762 men when its transport was sunk en route to the European continent by a German submarine—Roberts would later write a series of books about the disaster and its aftermath. (Courtesy of Ray Roberts.)

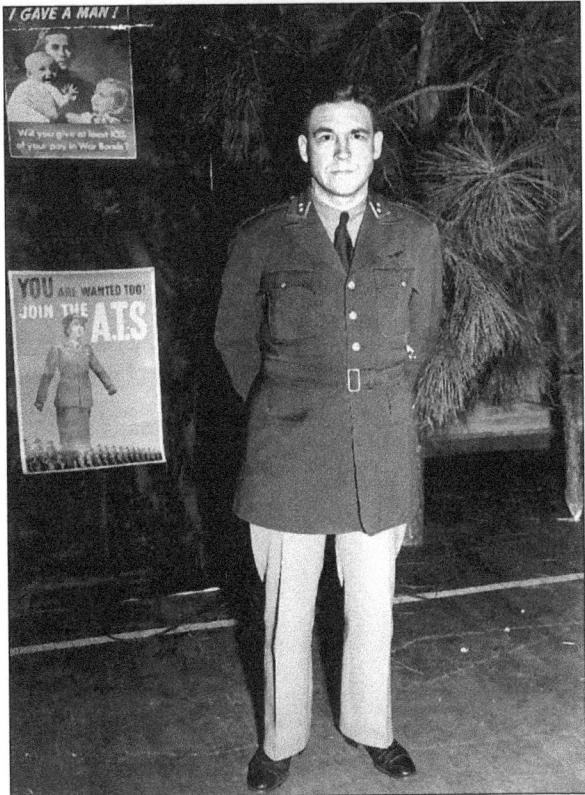

When Camp Rucker threw a United Nations Party during the 1942 Christmas season, the Netherlands dispatched Lt. Col. Conrad Giebel as the Netherlands' ambassador's special representative to the festivities. During the time, the Netherlands was still occupied by Nazi Germany. (Courtesy of AAM.)

A group of Camp Rucker officers, with Brigadier General Manley in the center, poses with Dutch and Soviet representatives at the camp's 1942 United Nations Party. "The United States," *Army Life* explained, "has made itself a part of a world-wide fighting organization, and now you, too, have made yourself a part of it. This worldwide group is known as the United Nations. Some of these nations contribute more directly on a larger scale than others, but that is less important than the fact that all are pitching into the fight as best they can." (Courtesy of AAM.)

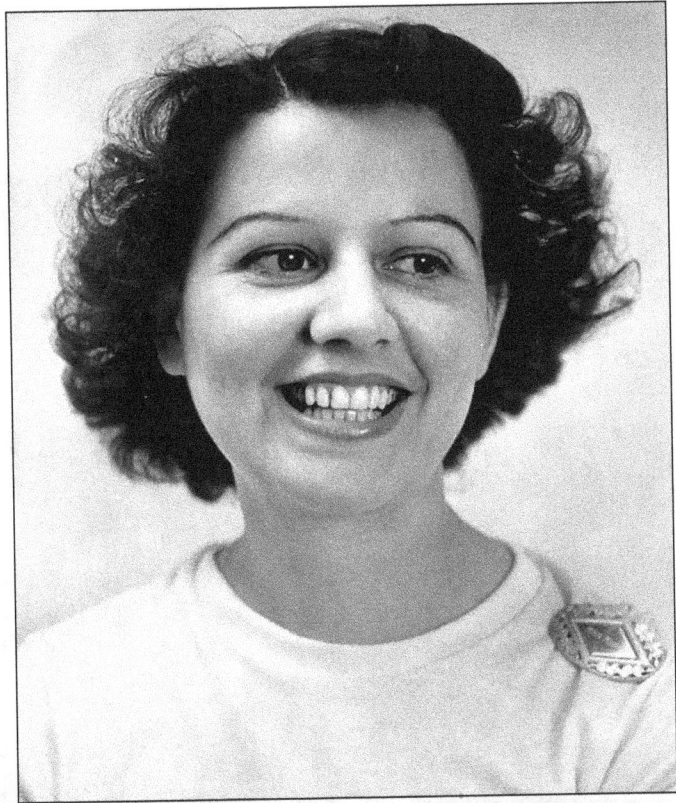

Miss Lillian H. Moore, shown here in June of 1943, was the hostess at Camp Rucker's Service Club No. 1. (Photo by U.S. Army Signal Corps, courtesy of AAM.)

This group of civilian volunteers and happy soldiers is gathered for the opening of Dothan's USO club in January of 1943. "The United Service Organizations provides your service club in town," explained *Army Life*. "It operates recreation, club, game and shower rooms; sleeping quarters; information and travel bureaus; dances; libraries; study classes; rest rooms and free checking in railroad stations and club houses, and motion pictures." (Courtesy of AAM.)

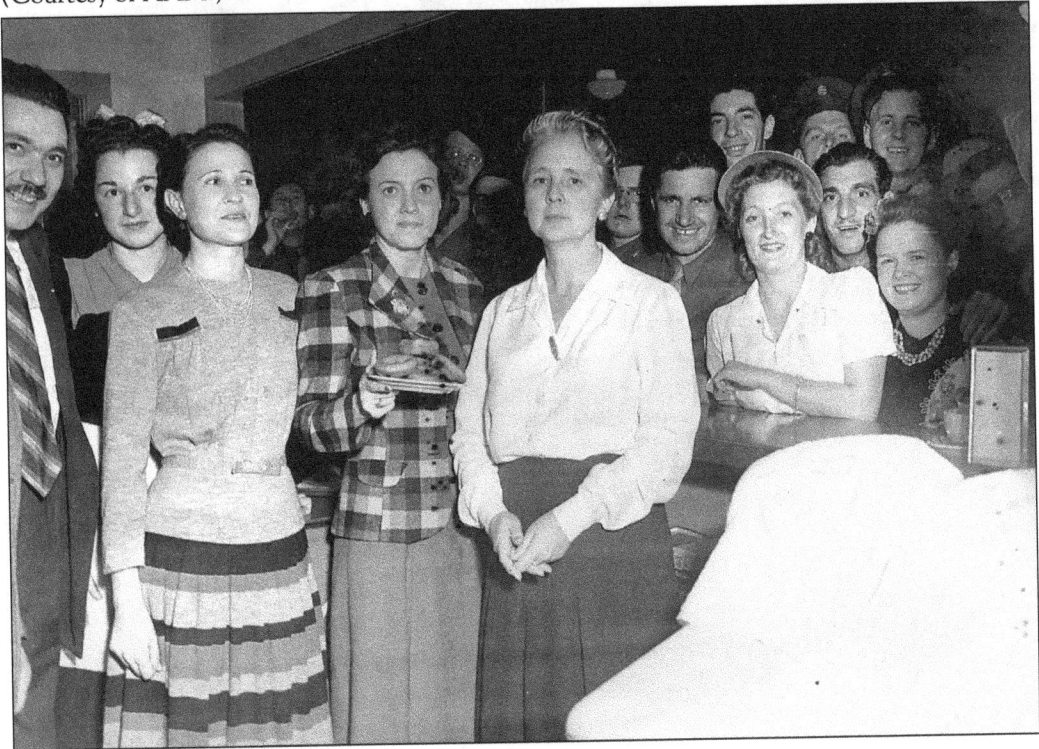

Soldiers and civilian hostesses pose for a photograph at the January 1943 opening of the USO club in Dothan. The USO, *Army Life* explained, "unites the efforts of many groups which have dedicated themselves to your welfare. Don't hesitate to use these services. They are maintained especially for you." (Courtesy of AAM.)

Five

SOLDIERING

These sergeants, shown in August of 1943, formed the cadre of the 204th Field Artillery Group. All veterans of overseas service, they had returned to the United States to help train the next wave of American artillerymen. "Fighting is now your business," *Army Life* warned. "It is a strange one to most of us. Things we don't understand are disturbing, so you may be disturbed about whether you will be ready and able to fight when you have to. . . . There is one factor more important than any other in overcoming fear. It is training." (Photo by U.S. Army Signal Corps, courtesy of AAM.)

General Manley presents Pvt. Effie Chism of Camp Rucker's WAC detachment with the Purple Heart Medal awarded posthumously to her husband. He had been killed in combat overseas in 1943. (Photo by U.S. Army Signal Corps, courtesy of AAM.)

This group of women, after completing their nurses training at Birmingham's Baptist Hospital, attended the Army Nurses Corps Basic Training Center at Camp Rucker. Once they graduated, they became lieutenants in the Nurses Corps and were qualified for service either in stateside military hospitals or in field units overseas. From left to right are Ruth Searcy, Anne Danna, Mattie Armstrong, Elsie F. Easterling, Ruby C. Nance, Grace Lynch, and Nancy F. Davenport. The Army had less than 700 nurses in uniform in 1939. By July of 1943, that number had jumped to over 36,000. (Photo by U.S. Army Signal Corps, courtesy of AAM.)

This 1943 photograph shows another group of recent Army Nurses Corps Basic Training Center graduates. These new lieutenants had originally completed their initial nurses training at Spartanburg General Hospital in Spartanburg, South Carolina. From left to right are Cartha Martie Bartless, Dorothy Louis Cothron, Gladys L. Burgess, Vera M. Drummond, Sybil E. Grant, Annie Marie Poole, and Sarah N. Hamilton. (Photo by U.S. Army Signal Corps, courtesy of AAM.)

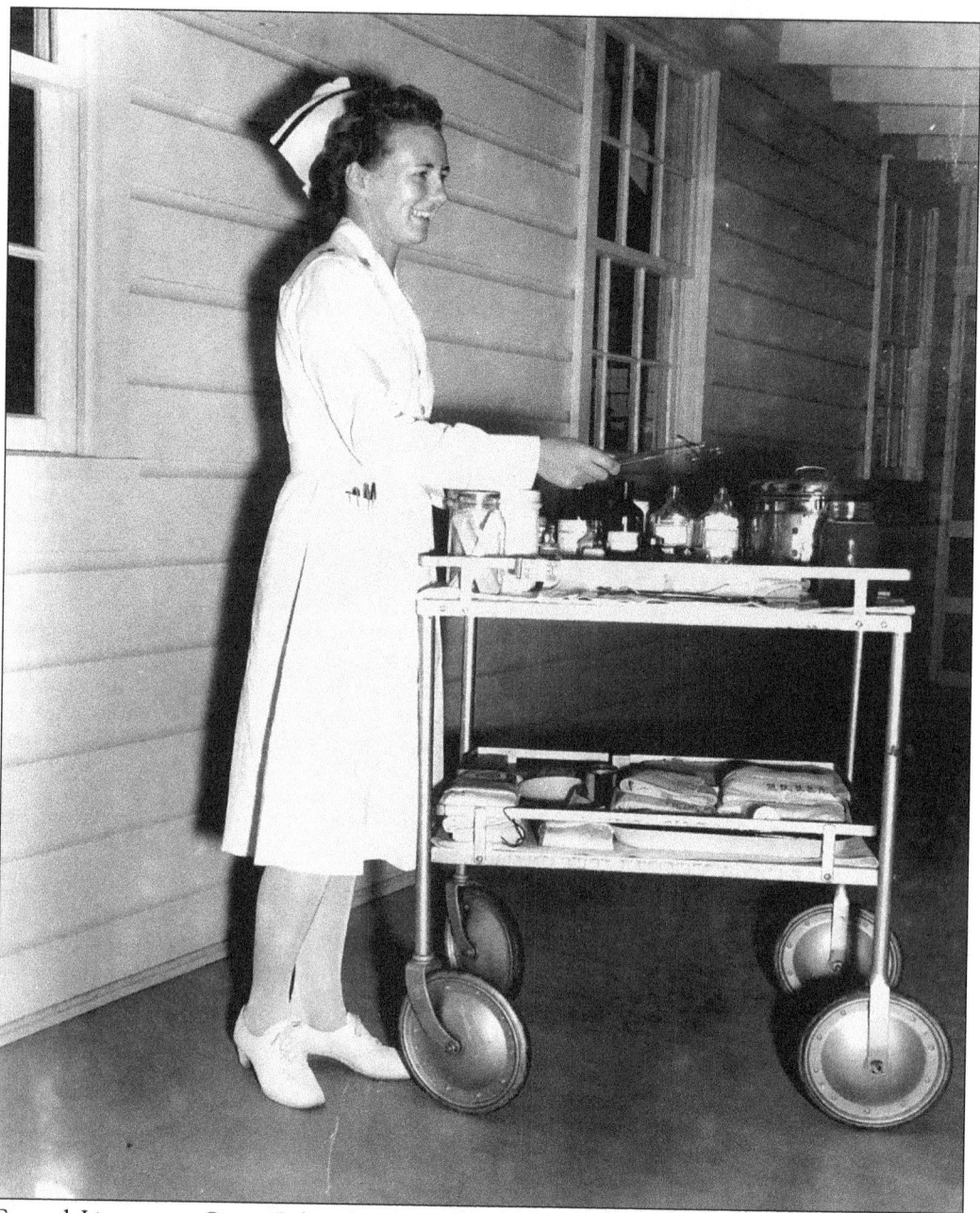

Second Lieutenant Grace Bahrenburg, a nurse with Camp Rucker's 3rd General Hospital, prepares to make her rounds in March of 1943. (Courtesy of AAM.)

Pfc. Helen Corenetz, shown here in August of 1943, was a WAC telegraph operator. "Perhaps soldiers will always demand the privilege of poking fun at the WACs, as big brothers deride their kid sisters," *Army Life* opined, "but if you hear any outsider making unflattering remarks about these girls who are serving their country with you, remember that they're part of the family—and this family doesn't take foolishness from anyone!" (Photo by U.S. Army Signal Corps, courtesy of AAM.)

Col. H.E. Thornton, commanding officer of the Fifth Detachment, Army Service Forces, Second Army, presents a rifleman's trophy to Pvt. Daniel Adams of the 297th Combat Engineers Battalion in July of 1943. (Photo by U.S. Army Signal Corps, courtesy of AAM.)

An M3 Lee medium tank carefully negotiates a log bridge during maneuvers at Camp Rucker. Ironically, the British called their version of this tank the "Grant." (Courtesy of AAM.)

Commanding officers at July 1943's Army Service Forces Review, with their unit guidon bearers behind them, prepare to report their units as present. In 1943, the Army was divided into three commands—the Army Ground Forces, the Army Air Forces, and the Army Service Forces. The Army Service Forces encompassed such diverse technical services and staff divisions as the Quartermaster Corps, Ordnance Corps, Corps of Engineers, Chemical Warfare Service, Signal Corps, Medical Department, Transportation Corps, Provost Marshal, Chaplains, and the Adjutant General and Judge Advocate General Corps. (Photo by U.S. Army Signal Corps, courtesy of AAM.)

Camp Rucker, like all Army camps, depended upon a complement of Military Police, or MPs, to maintain order and to provide camp security. This photograph of Camp Rucker's MPs was

taken in April of 1943. (Courtesy of AAM.)

A detachment of nurses from Camp Rucker's hospital parades through the streets of Dothan on October 2, 1943, en route to watch the premiere of the film *This is the Army*. (Photo by U.S. Army Signal Corps, courtesy of AAM.)

Led by 1st Lt. Agnes J. Peterson, a detachment of WACs marches toward the reviewing stand during the Army Service Forces review in July of 1943. (Photo by U.S. Army Signal Corps, courtesy of AAM.)

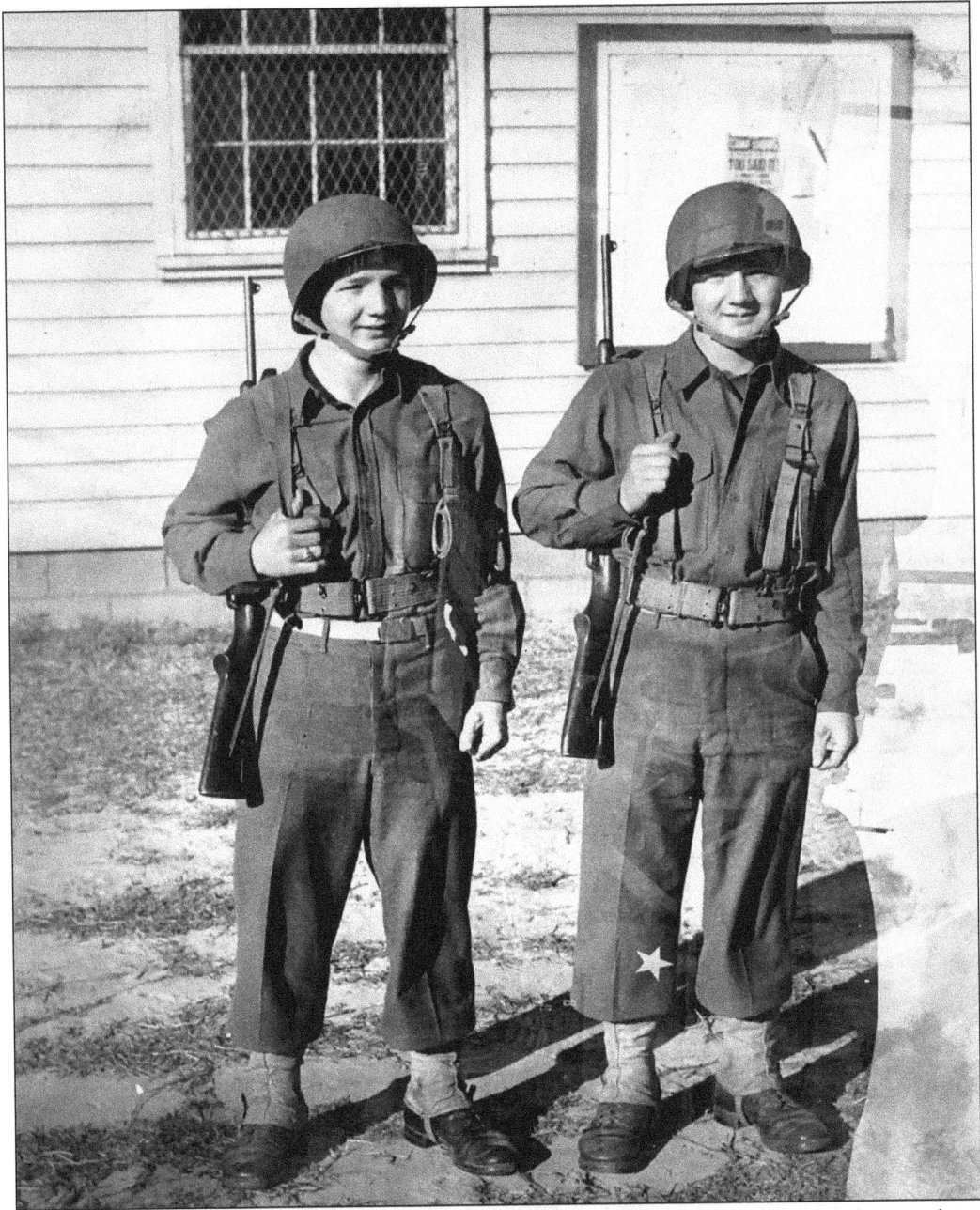

Pvts. Herbert and Harvey Smith, identical twins from Gettysburg, Pennsylvania, left Gettysburg College in February of 1943 to enlist. At Camp Rucker, they were assigned to the same field artillery unit. The unit's officers reportedly simplified matters by always assigning them to the same guard or work details together—otherwise, they wouldn't know which Smith was reporting for duty. (Photo by U.S. Army Signal Corps, courtesy of AAM.)

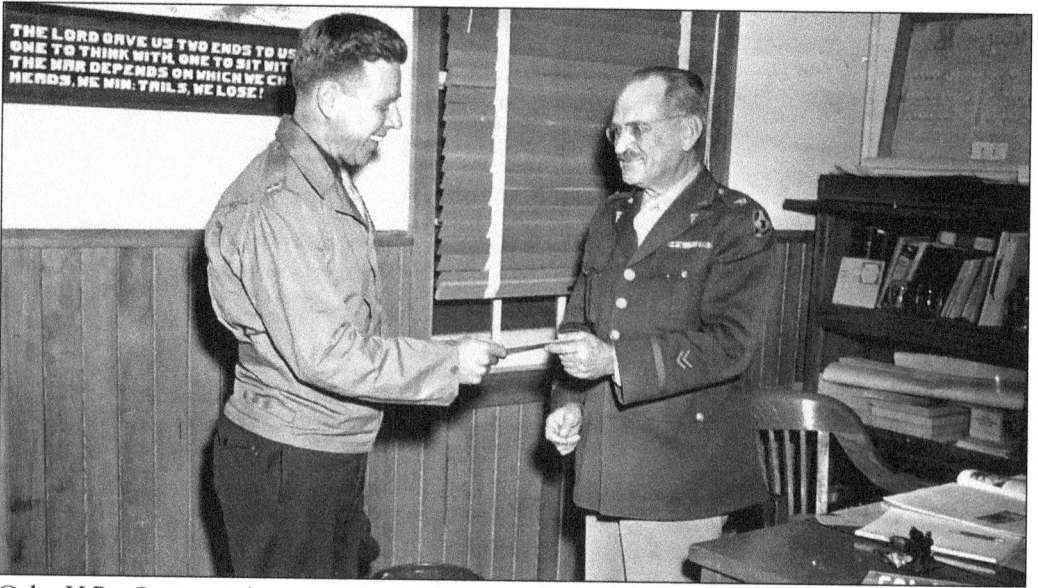

Col. H.P. Sawyer, the commanding officer of Camp Rucker's hospital, presents a commendation to Lt. William S. Cooney Jr., of Bridgeport, Connecticut. The commendation honored Cooney's efforts in the South Pacific theater of operations in October of 1943. When the transport ship on which he was sailing was torpedoed, Cooney (then a sergeant) remained aboard the sinking vessel until all wounded personnel had been lowered safely into the lifeboats. (Photo by U.S. Army Signal Corps, courtesy of AAM.)

As part of a War Bonds drive in southeastern Alabama, the Army offered rides in this jeep chauffeured by WAC corporal Madeline D. "Tony" Antonioli to civilians purchasing War Bonds. The drive was wildly successful—the government raised $95,000 to support the war effort. (Photo by U.S. Army Signal Corps, courtesy of AAM.)

The chemical warfare companies that trained at Camp Rucker relied on a variety of specialized equipment foreign to other units' inventories. This photograph shows the direction and speed of wind on a battlefield. Such information would be critical in ensuring the chemical company's placement of accurate and dense smoke screens to conceal friendly troop movements on the battlefield. (Photo by U.S. Army Signal Corps, courtesy of AAM.)

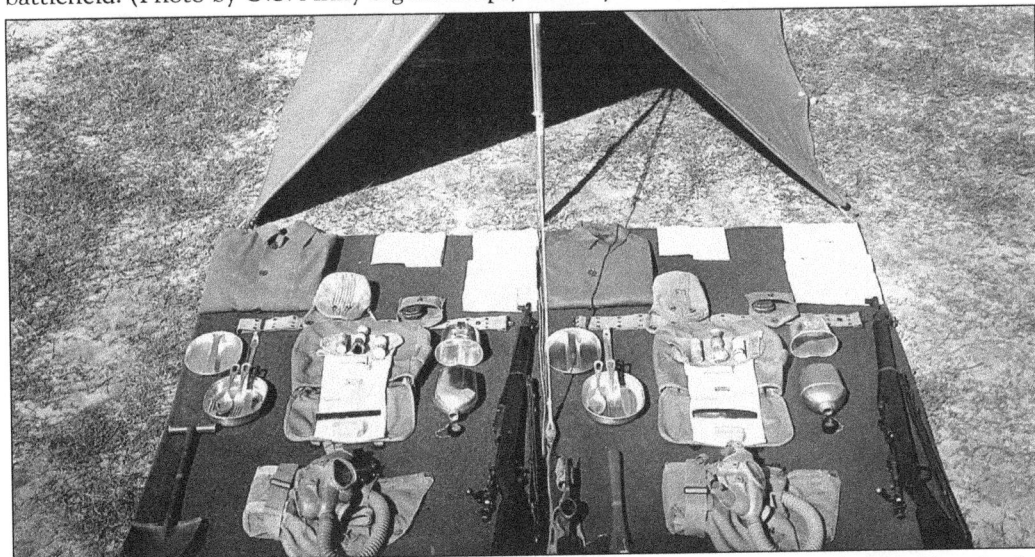

Here, two soldiers in a chemical warfare unit have set out the contents of their field packs for inspection in front of their shared pup tent. Their equipment ranged from toothbrushes and underwear to gas masks and M1903 .30 caliber Springfield rifles. "Your Government Issue equipment is loaned to you for the duration: it belongs directly to the Government," *Army Life* stated. "If you are negligent and your equipment is lost, damaged or destroyed through your own carelessness, you will have to pay for it." (Photo by U.S. Army Signal Corps, courtesy of AAM.)

This photograph displays the wide array of communications equipment—radios, field telephones, and switchboards—upon which a chemical warfare battalion would depend in order to stay in communication with other units. Although equipped for its use, the United States Army never used poison gas during World War II. Instead, chemical warfare units were used to lay smoke screens and to provide conventional fire support. (Photo by U.S. Army Signal Corps, courtesy of AAM.)

Under the watchful eyes of a lieutenant and one of their sergeants, the men of the 37th Medical Ambulance Battalion learn how to erect a canvas tent. "You must," lectured *Army Life*, "take your training seriously. Training will make you—as an individual—able to win *your* fights. Give it your best. It you don't, you have the most to lose." (Photo by U.S. Army Signal Corps, courtesy of AAM.)

A group of teenaged privates of the 736th Tank Battalion pose in front of one of their unit's M4 Sherman tanks in March of 1943. From left to right are Calvin H. White, Leland McFarren, Ray Blake Jr., Victor Saucerman, Joe E. Perry, and Anthony Hrasack. By the end of the war in Europe, this battalion's tanks had crossed the Elbe River and were reportedly the closest Shermans to the Nazi capital of Berlin. (Courtesy of AAM.)

Lt. Col. William M. Hernandez, commanding officer of the 628th Tank Destroyer Battalion, stands in front of one of his unit's M10 tank destroyers in October of 1943. The 628th was formed from field artillery units of the Pennsylvania National Guard's 28th Infantry Division and arrived in Camp Rucker in May of 1943 following amphibious training in the Gulf of Mexico. The battalion reached France on July 30, 1944, and proceeded to fight its way across northwestern Europe before ending the war on the Elbe River. Lieutenant Colonel Hernandez, however, was killed in action in Douains, France, on August 20, 1944, while directing indirect fire on enemy tanks near the Seine River. (Photo by U.S. Army Signal Corps, courtesy of AAM.)

Lt. Col. John White, a 1932 graduate of Purdue University (where he had been a star quarterback on the university's football team), was one of the officers tasked with supervising the training of young tankers at Camp Rucker. He is shown here in the winter of 1943 in the turret of an M3 light tank. (Photo by U.S. Army Signal Corps, courtesy of AAM.)

Col. Edward A. Kimball, commanding officer of Camp Rucker's 8th Tank Group, stands in front of a M3 Lee medium tank in November of 1942. Although variants of the M3 were used in large numbers by British forces in Libya and Egypt in 1942, the United States used it primarily as a training vehicle after some initial combat use in North Africa and the Pacific. (Courtesy of AAM.)

Soldiers of the 204th Field Artillery Group receive Good Conduct Medals during a unit review in January of 1944. "If you're the right kind of soldier," predicted *Army Life*, "you will very soon take a great deal of pride in the way you wear your uniform. You will soon notice that the neat,

well-pressed soldier is the one who is respected and receives the best treatment." (Photo by U.S. Army Signal Corps, courtesy of AAM.)

A unit of Army nurses practices erecting two-man pup tents. "Much of your time in the Army will be spent outdoors," *Army Life* warned, "and except for certain phases of instruction, weather conditions will not alter training." (Photo by U.S. Army Signal Corps, courtesy of AAM.)

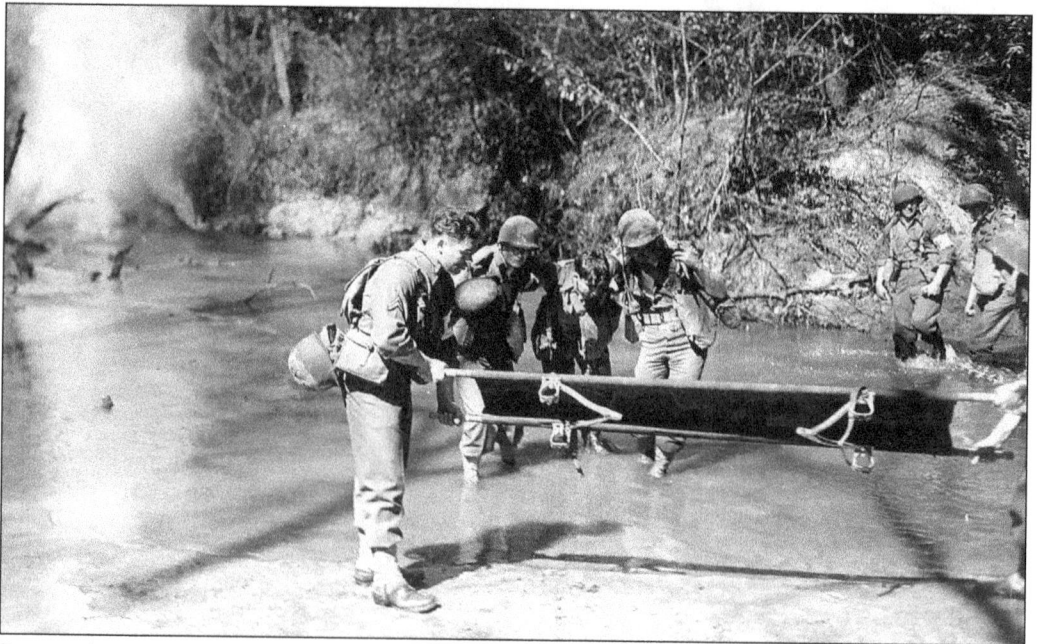

During this training exercise in March of 1943, a litter team from the 37th Medical Ambulance Battalion helps move a mock casualty to a waiting litter. "If you played on a football team in school you know that before your very first game you went through weeks of preparation," *Army Life* explained. "The final test of your success as a football team came in 'the big game.' In the Army, all your individual and group training must be put to the test, too. And the battle is the pay-off. The results are counted in victory or defeat." (Courtesy of AAM.)

A litter party of 37th Medical Ambulance Battalion soldiers carries a wounded comrade on a stretcher through a haze of smoke. They are wearing gas masks during this training exercise. "When everyone does all in his power for the success of the whole outfit, that's teamwork," *Army Life* declared. "If you and your mates pull together now in training, you will be a closely-knit, smoothly functioning unit later in combat." (Courtesy of AAM.)

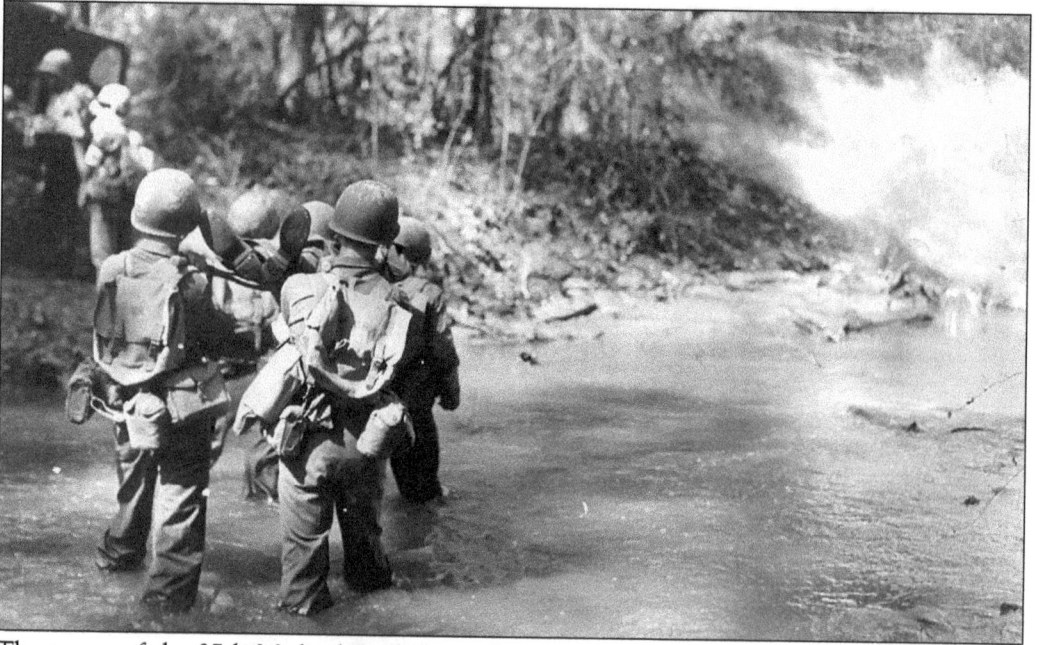

The troops of the 37th Medical Ambulance Battalion lug a comrade on a stretcher balanced on their shoulders as they head across a Camp Rucker creek for a waiting field ambulance. The battalion would eventually land in Europe in October of 1944 and would help evacuate wounded soldiers from battlefields in northern France, the Ardennes Forest, and Germany's Rhineland. (Courtesy of AAM.)

During a training exercise in March of 1943, a 37th Medical Ambulance Battalion litter team discovers a mock casualty sprawled across a fallen tree. Now they will have to carry him back to a field ambulance for evacuation to the rear for medical treatment. (Courtesy of AAM.)

Litter teams from the 37th Medical Ambulance Battalion evacuate casualties from a mock battlefield to a waiting ambulance. Explosives detonated in the creek behind them add to the exercise's stress. "Maneuvers are mock battles," *Army Life* lectured. "Many of the conditions with which you will be faced in actual warfare are dealt with on maneuvers. You will be made to sense what warfare is like in every respect except for one: Safety precautions will be taken to avoid casualties." (Courtesy of AAM.)

A tired team of 37th Medical Ambulance Battalion litter-carriers negotiates a fallen tree as they move along a creek during a training exercise. (Courtesy of AAM.)

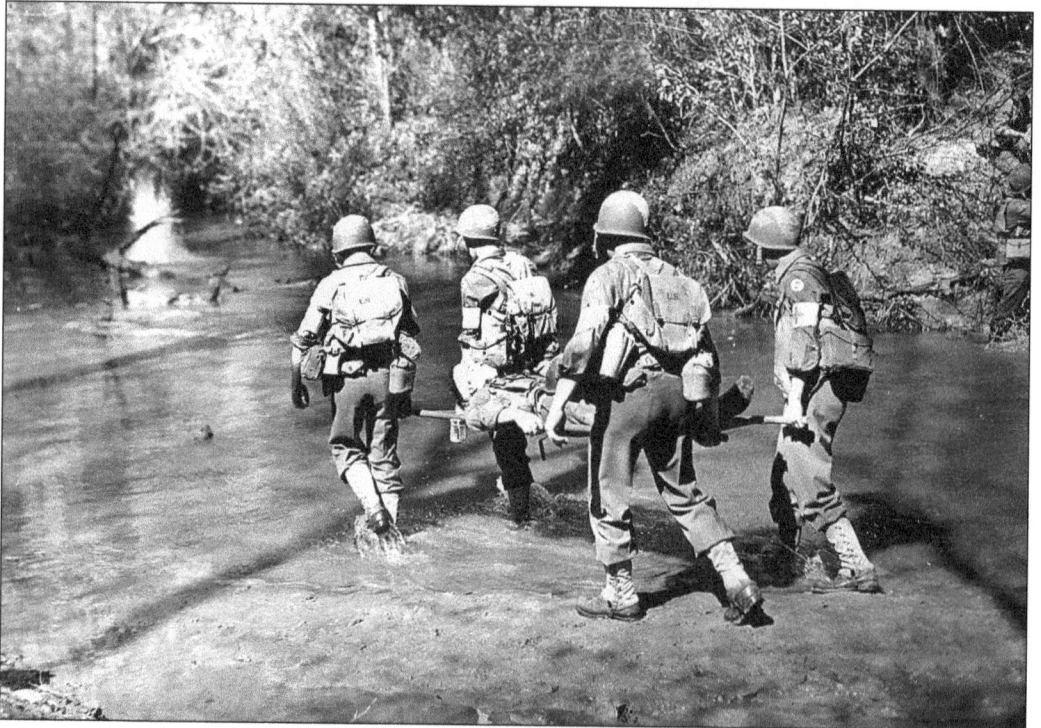

Soldiers of the 37th Medical Ambulance Battalion stagger wearily down a creek as they carry a stretcher-borne casualty. (Courtesy of AAM.)

An exploding shell in front of this 37th Medical Ambulance Battalion litter party causes the soldiers to stop short in their tracks. "Maneuvers are not only an experience to show you what it will be like in combat," *Army Life* explained. "They are also a test to determine whether you and your unit have had sufficient and proper training. Any unit that functions efficiently on maneuvers is a unit that should be a 'crack' outfit on the battlefront." (Courtesy of AAM.)

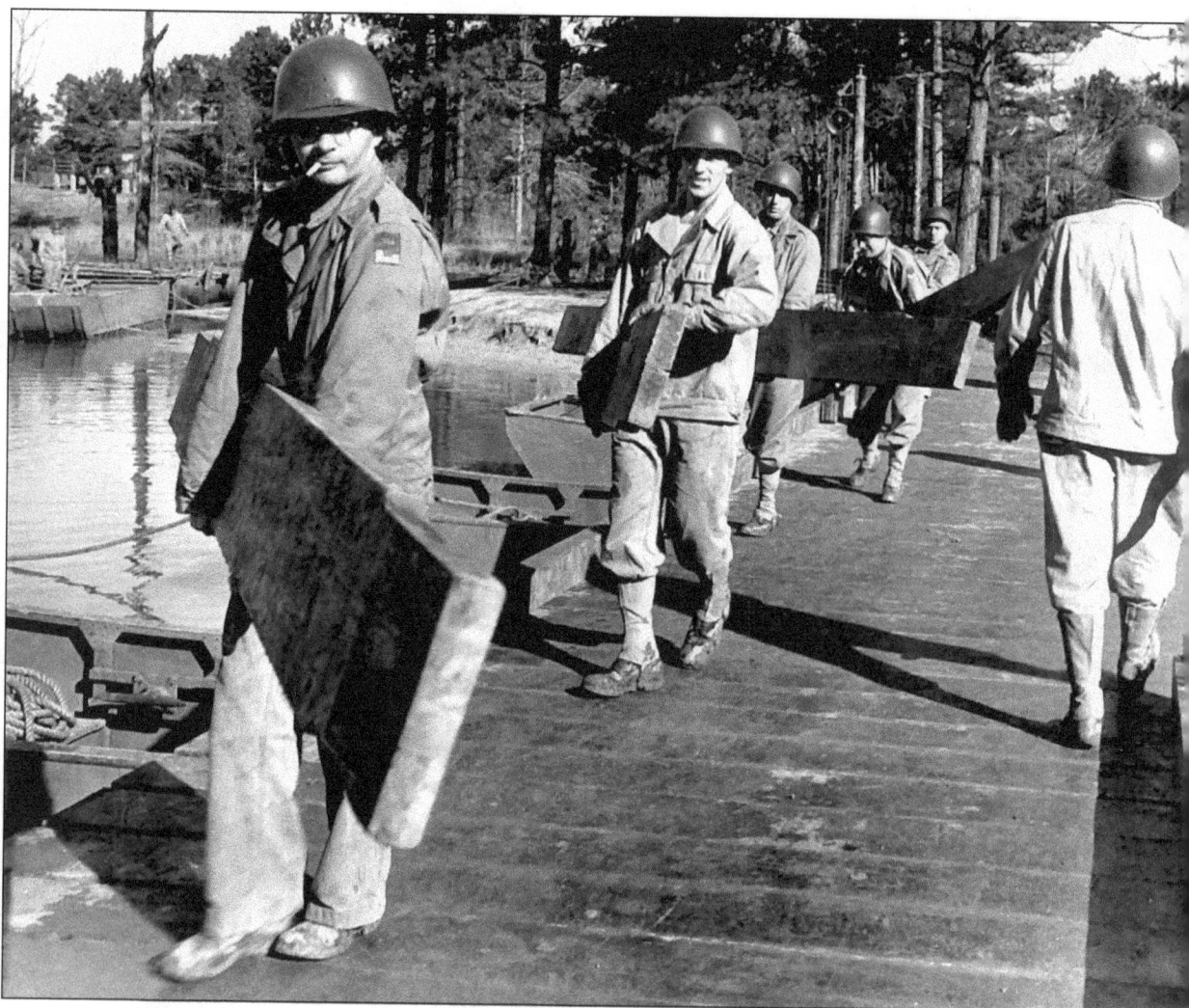

Soldiers of the 336th Combat Engineer Battalion lug planks across the completed span of a pontoon bridge they are constructing. "It's not possible to grasp and retain everything which you must learn in a short time," *Army Life* predicted. "Don't expect to look like a tough old-timer and to feel seasoned and assured in a week or month. . . . Don't fret because some things which are taught to you may seem contrary to everything you have ever known. You'll have to file away your civilian knowledge for future use." (Courtesy of AAM.)

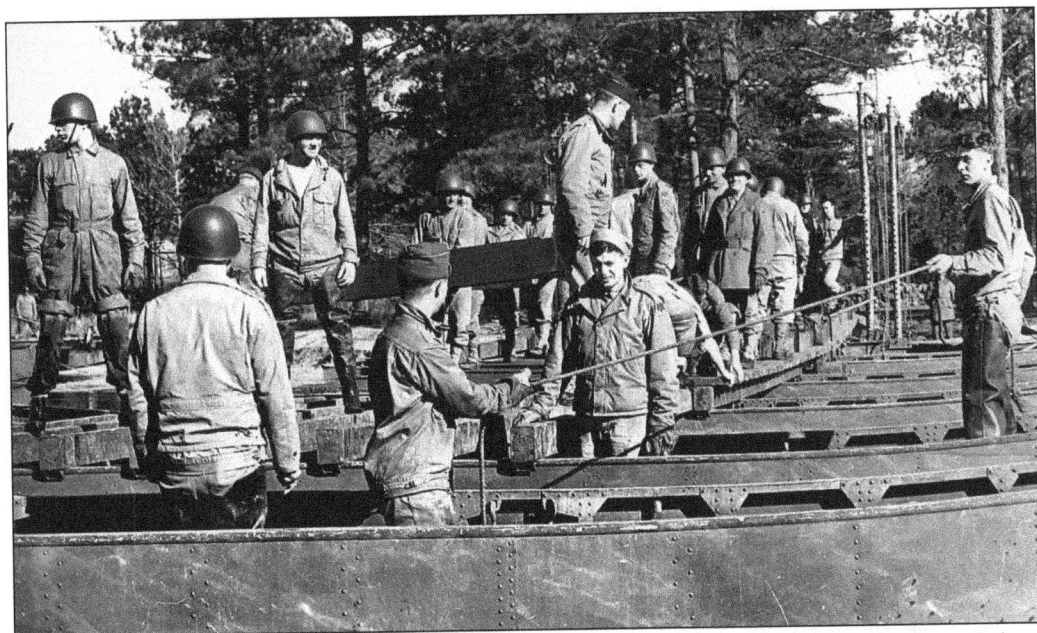

Combat engineers of the 336th Combat Engineer Battalion stand in the pontoons of their bridge, holding them steady and awaiting the arrival of more planks to complete the roadway atop the pontoons. (Courtesy of AAM.)

The 336th Combat Engineer Battalion hurriedly slaps a pontoon bridge together across a lake inlet. Bridges like this would be critical in supporting the Allied drive across Europe in the coming months. (Courtesy of AAM.)

A private of the 336th Combat Engineer Battalion pauses to flash the camera a tired grin while his unit labors to construct a pontoon bridge. "The Japs and Germans," *Army Life* lectured, "have been subjected to rigid military training for years while ours is an Army of civilians mustered into uniform. Yet ours is a potent Army because we Americans have a tremendous ability to learn speedily as individuals and to act cooperatively as members of teams." (Courtesy of AAM.)

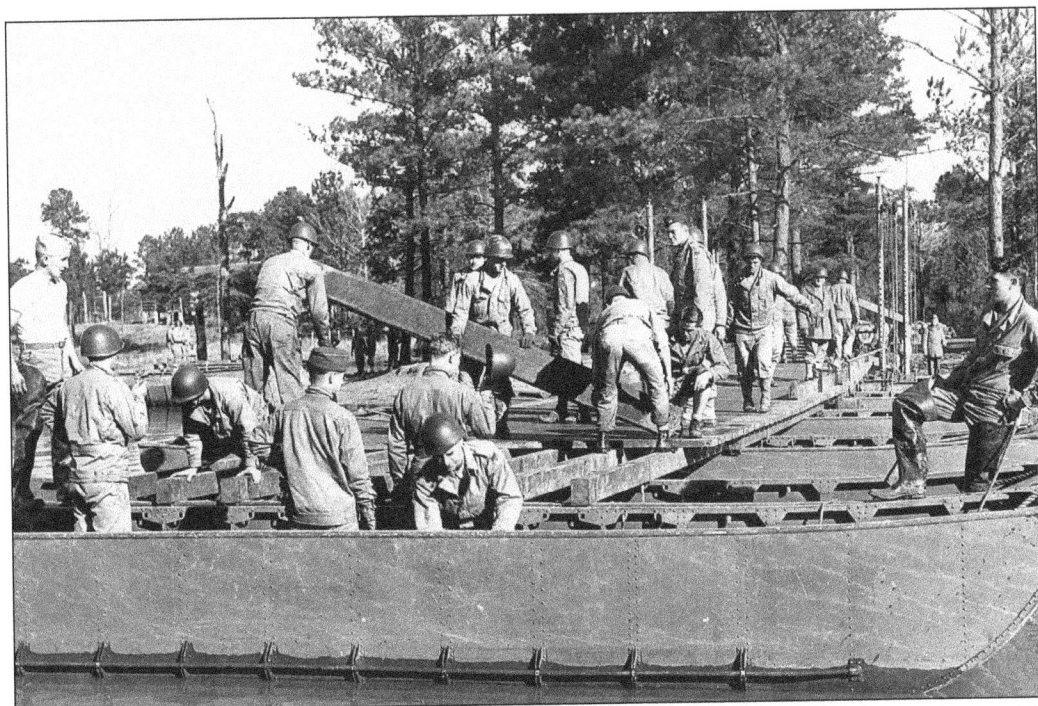

The engineers' pontoon bridge nears completion as the last planks are brought in to span the remaining pontoons. (Courtesy of AAM.)

Men of the 28th Ordnance Company construct an obstacle course near a set of Camp Rucker's barracks. (Courtesy of AAM.)

A firing line of soldiers is arrayed up range from their targets at one of Camp Rucker's firing ranges. (Photo by U.S. Army Signal Corps, courtesy of AAM.)

Soldiers practice their marksmanship from the prone firing position during weapons training. "Very soon," *Army Life* stated, "you will have an opportunity to qualify as a gunner with one or more types of weapons. For qualifying as an expert gunner, sharpshooter (1st class gunner) or marksman (2nd class gunner), you will be awarded a badge which you will wear on your blouse." (Photo by U.S. Army Signal Corps, courtesy of AAM.)

These Camp Rucker soldiers, pictured at a firing range in January of 1944, are armed with .30 caliber M1 carbines. "Many men who have had little experience with firearms are now in the Army as recruits," *Army Life* warned. "Sooner or later, each of them must handle weapons for the first time. Whether you are an old hand or a novice at weapons-handling, *handle them with care!* These weapons are made to kill." (Photo by U.S. Army Signal Corps, courtesy of AAM.)

This photograph shows the behind-the-scenes work at one of Camp Rucker's firing ranges. In the concrete-protected firing pits, teams of two soldiers man each target. One raises, lowers, and scores each target for the shooter up range, while the other coordinates with the firing line by field telephone. (Photo by U.S. Army Signal Corps, courtesy of AAM.)

Troops manning the targets at one of Camp Rucker's fire ranges pause during a break in the training. (Photo by U.S. Army Signal Corps, courtesy of AAM.)

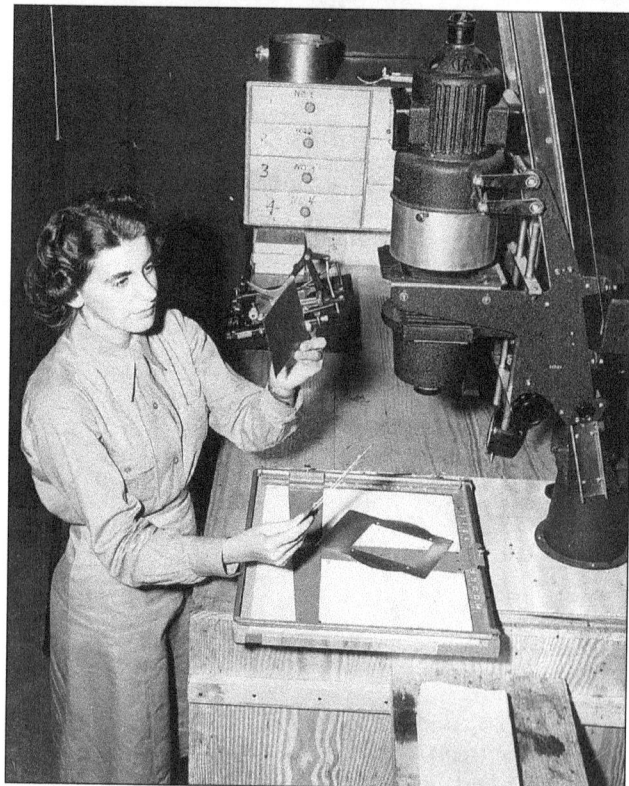

Pvt. Eleanor Zangrilli, a WAC darkroom technician, examines a photographic print in a Camp Rucker darkroom in August of 1943. (Photo by U.S. Army Signal Corps, courtesy of AAM.)

Soldiers of the 851st Ordnance Company work to repair one of Camp Rucker's jeeps. (Photo by U.S. Army Signal Corps, courtesy of AAM.)

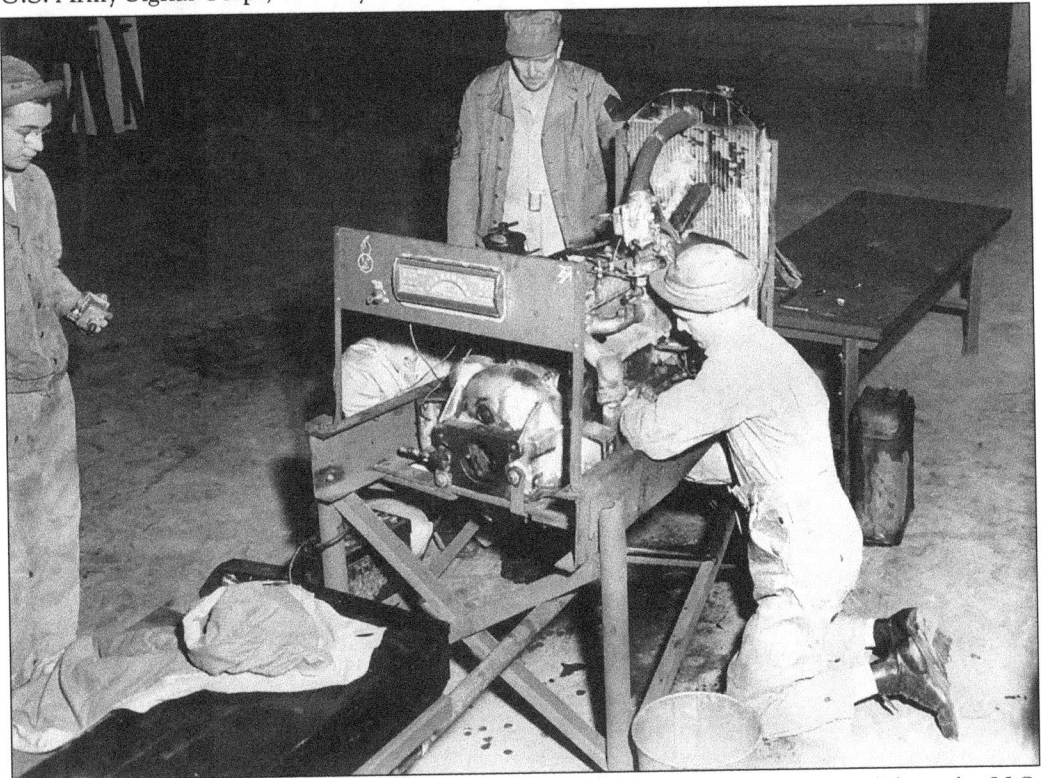

In this photograph, 851st Ordnance Company mechanics tune up a motor. (Photo by U.S. Army Signal Corps, courtesy of AAM.)

Soldiers of one of Camp Rucker's Chemical Warfare Service units carefully load a 900-gallon drum of flamethrower fuel into a trailer. (Photo by U.S. Army Signal Corps, courtesy of AAM.)

A medical team at a first aid station splints a mock casualty during a field exercise. (Photo by U.S. Army Signal Corps, courtesy of AAM.)

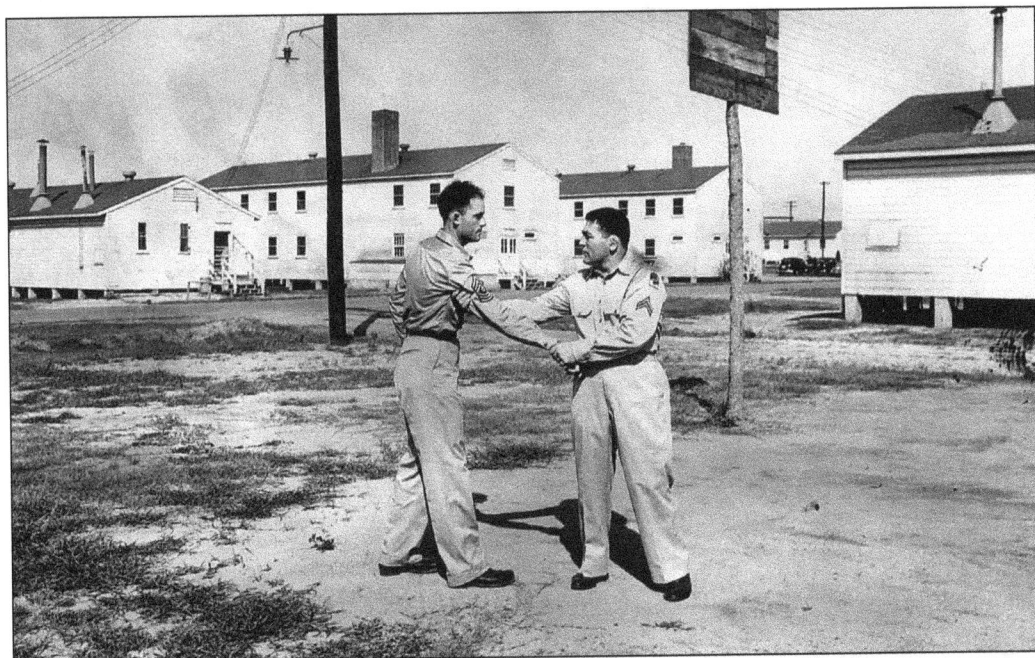

Cpl. Nicholas McRadu prepares to demonstrate a judo move to 1st Sgt. Billy M. Fann in July of 1943. Both men were assigned to the 87th Chemical Battalion. (Photo by U.S. Army Signal Corps, courtesy of AAM.)

With a quick twist, Corporal McRadu tosses his first sergeant over his shoulder during an exhibition of his judo skills. "This is the real thing. You must set your mind to it, clear your head for combat action," *Army Life* advised. "You'll have to forget your civilian life, your best girl, your family; for thinking about what you might lose makes it harder to concentrate on winning." (Photo by U.S. Army Signal Corps, courtesy of AAM.)

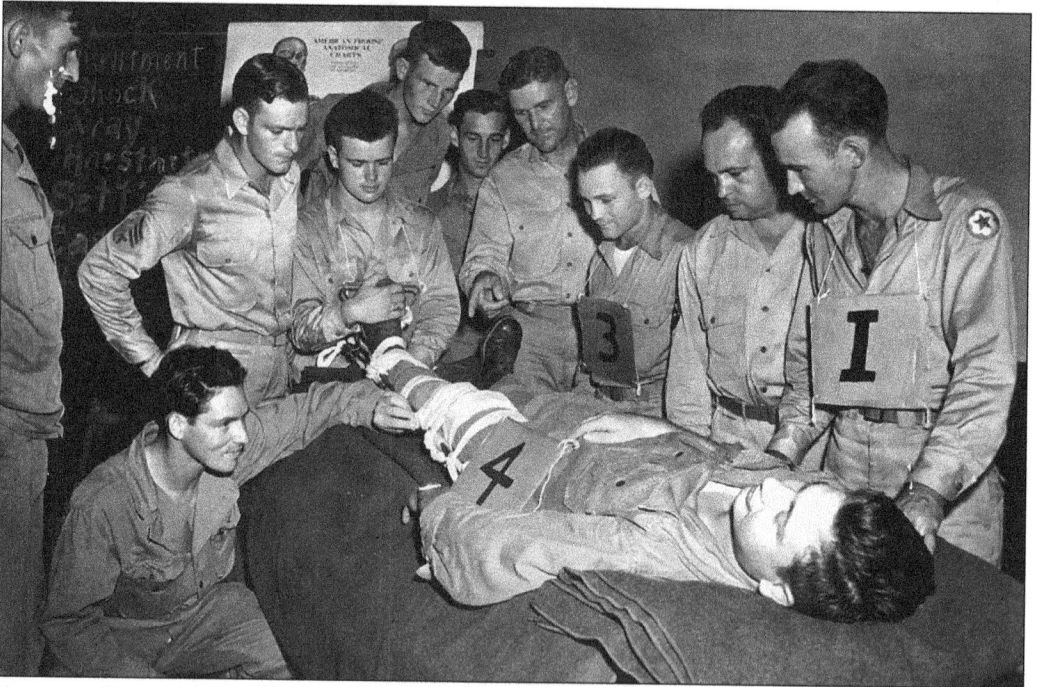

A team of would-be medics evaluates a mock casualty under the watchful eye of a technical sergeant. Keep a casualty warm and calm, *Army Life* recommended. "Loosen clothing to make breathing easy. Stop bleeding by the best means available. Get a medical officer or an enlisted man of the Medical Corps as quickly as possible." (Photo by U.S. Army Signal Corps, courtesy of AAM.)

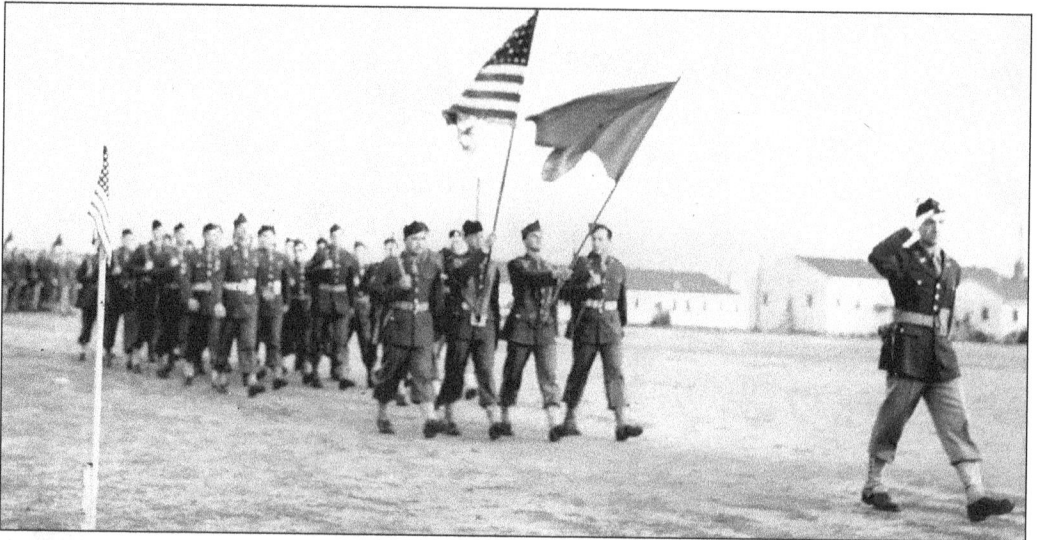

A unit of the 204th Field Artillery Group passes in review during a parade in January of 1944. "Good soldiers," *Army Life* explained, "have a characteristic manner of walking. Part of this is due to the rate at which they walk. This rate, known as 'cadence,' is 120 steps per minute, or 2 steps per second. This is the cadence of 'quick time.' All steps and facings, as well as the manual of arms, are executed normally at 'quick time.'" (Photo by U.S. Army Signal Corps, courtesy of AAM.)

A group of soldiers lines up for payday disbursements in the field. Note the drawn sidearms of the paymasters seated at the table. "You can be practically certain that you will receive enough actual cash to take care of all essentials, and that you'll have some left over for extras to make life more pleasant," *Army Life* assured its readers. (Photo by U.S. Army Signal Corps, courtesy of AAM.)

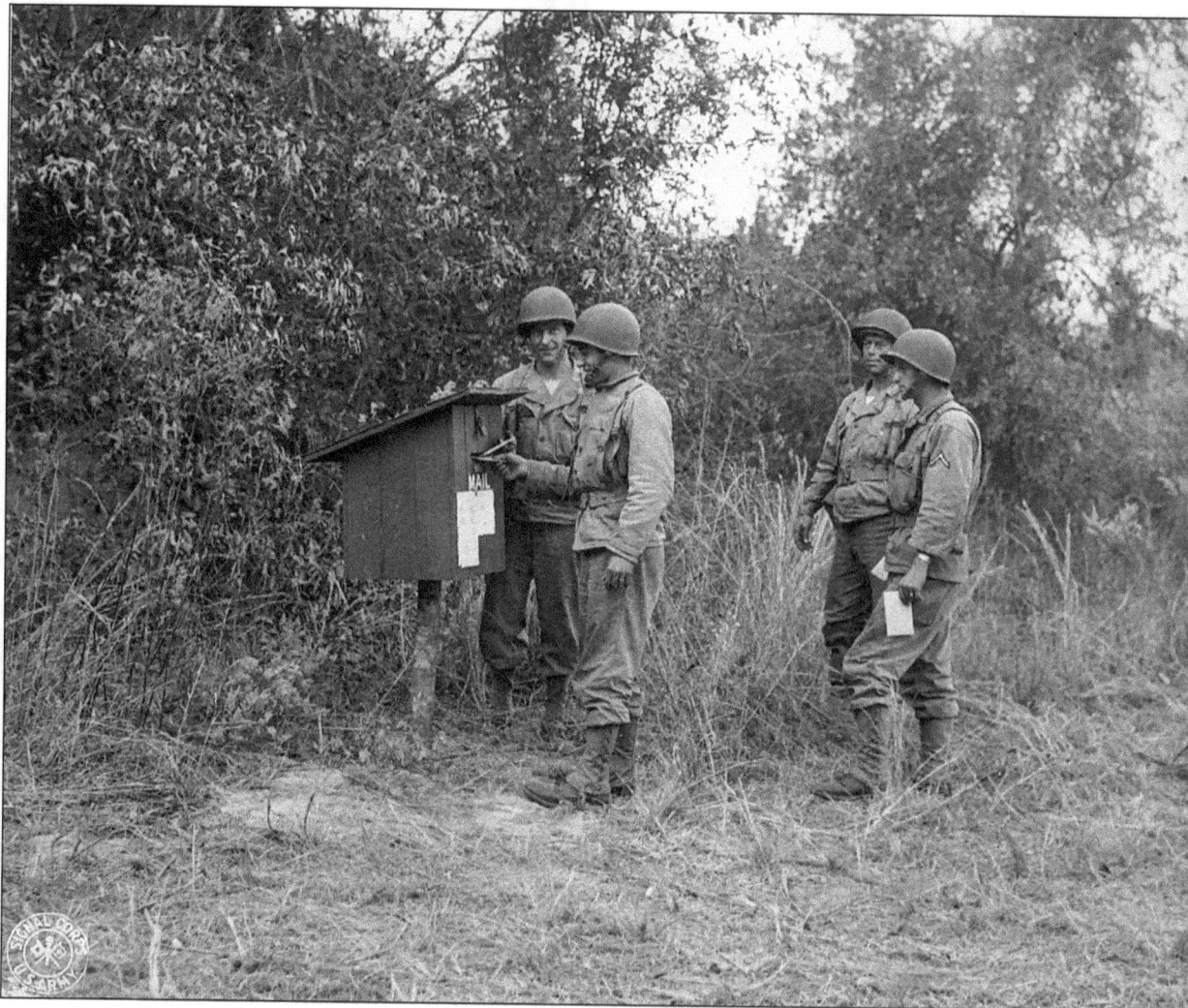

Camp Rucker soldiers drop off letters to loved ones back home at a makeshift mailbox in the field. "Tell your folks at home the truth about your life," *Army Life* encouraged. "[D]on't dwell on the hard side of things nor the uncertainties of the future. They'll be interested in your accomplishments, your work, your fun. Give them a chance to be proud of their soldier." (Photo by U.S. Army Signal Corps, courtesy of AAM.)

Pfc. Dorothy Olson passes through the chow line at the WAC's mess hall. "It will be a great day," *Army Life* predicted, "when you have to admit (probably only to yourself!) that Army chow isn't the beans, canned willie and slumgullion that your father about in World War I; that it's plenty good eating and—if it weren't for KP!—you wouldn't have a squawk in the world!" (Courtesy of AAM.)

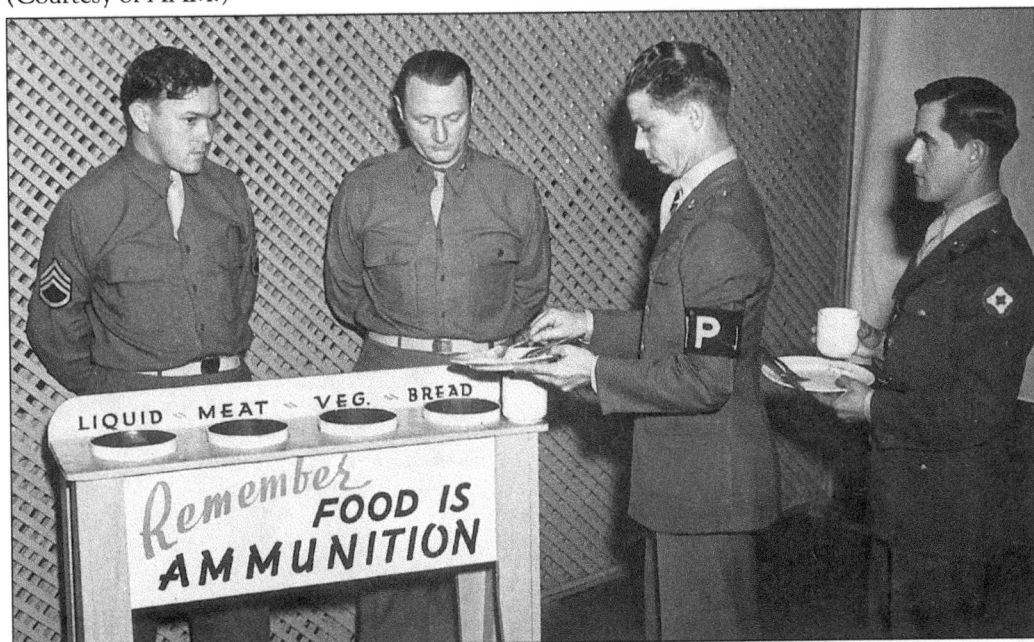

A military policeman scrapes off his plate at one of Camp Rucker's mess halls. A mess officer and mess sergeant keep a watchful eye over the line to make sure no one is throwing away too much food. "If you do not already realize it," *Army Life* warned, "you will soon learn that the Army is scrupulously careful to control consumption and eliminate waste. This care is particularly close with regard to food." (Photo by U.S. Army Signal Corps, courtesy of AAM.)

Col. William Walters, commanding officer of the 204th Field Artillery Group, explains a display of his battalion's equipment to a group of newspapermen and radio reporters during the 1944 War Bonds Drive. (Photograph by U.S. Army Signal Corps, courtesy of AAM.)

Medics fill out prescriptions in the pharmacy of the Camp Rucker hospital during April of 1943. (Courtesy of AAM.)

Members of the 336th Engineers' "Drum and Bugle Corps" pose with their instruments during a break in the battalion's training. (Courtesy of AAM.)

Soldiers of the 336th Combat Engineer Battalion pull a cable across a set of girders on a bridge they are erecting over Camp Rucker's Claybank Creek during a bridge building exercise. "Your training in garrison and on maneuvers will be so complete," *Army Life* assured its readers, "that you not only will understand all of it, but that when you are put to the battle test, your training will come back to you instinctively." (Courtesy of AAM.)

A group of 336th Combat Engineer Battalion soldiers watch the construction of a bridge from the banks of Camp Rucker's Claybank Creek. (Courtesy of AAM.)

Once the 336th's Combat Engineer Battalion's bridge was complete, it was time for its first customer—an M3 Lee medium tank of the 8th Tank Group. (Courtesy of AAM.)

The M3 Lee medium tank's crew poses on their vehicle atop the engineers' bridge. The standing tanker is resting his foot on the tank's secondary armament, a 37 mm gun. The vehicle's main gun, placed in a sponson in the tank's hull, was a more lethal 75 mm. Eventually, the M3 Lee would be replaced by the M4 Sherman. It continued to serve, however, as a training tank at stateside Army posts. (Courtesy of AAM.)

This photograph shows a front view of the M3 Lee medium tank. The 60,000-pound tank carried a cramped six-man crew; two of them are visible in this image. (Courtesy of AAM.)

Soldiers of the 336th Combat Engineer Battalion pick their way across a rope bridge on an obstacle course at Camp Rucker. (Courtesy of AAM.)

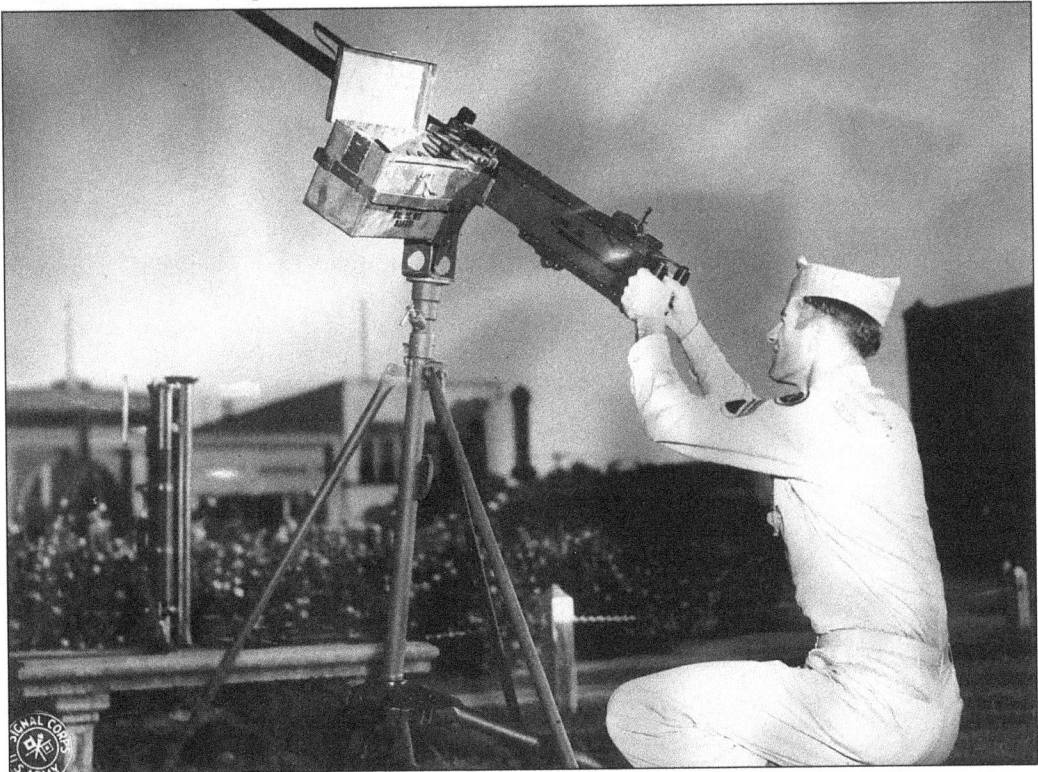

A sergeant fends off a mock air attack with his .50 caliber machine gun during a rally to encourage textile workers in the south Alabama town of Andalusia. (Courtesy of AAM.)

Following the Allied victory in Europe, happy veterans of the 766th Ordnance Company climb up a gangplank to board a troopship for the voyage back to the States. The 766th Ordnance Company was one of the dozens of units that had prepared for war in Europe and the Pacific by training at Camp Rucker. "We are a peaceful people," declared *Army Life*. "In a world of might and conquest, we tried to maintain good relations, but we were forced to fight. . . . America does not start wars, but it finishes them." (Photograph courtesy of Ray Roberts.)

Visit us at
arcadiapublishing.com

www.ingramcontent.com/pod-product-compliance
Lightning Source LLC
Chambersburg PA
CBHW080605110426
42813CB00006B/1413